DEVELOPMENTS IN NEAR-INFRARED SPECTROSCOPY

Edited by **Konstantinos G. Kyprianidis**
and **Jan Skvaril**

Developments in Near-Infrared Spectroscopy

http://dx.doi.org/10.5772/62932
Edited by Konstantinos G. Kyprianidis and Jan Skvaril

Contributors

Mohamad Sawan, Amal Kassab, Mercedes Lopez, Marie-Jeanne Bertrand, Philippe L. L'Allier, Jean-Claude Tardif, Iwao Mizumoto, Luis Galvez -Sola, Juan J. Martinez-Nicolas, Raquel Muelas-Domingo, Manuel Nieves, Francisco Garcia-Sanchez, Philiswa Nomngongo

CBS, Edition **2018**

Published by InTech

Janeza Trdine 9, 51000 Rijeka, Croatia

Publishing Process Manager Romina Rovan

Technical Editor SPi Global

Cover Designer InTech Design Team

Additional hard copies can be obtained from orders@intechopen.com

Developments in Near-Infrared Spectroscopy, Edited by Konstantinos G. Kyprianidis and Jan Skvaril
p. cm.
Print ISBN 978-953-51-3017-8
Online ISBN 978-953-51-3018-5

Contents

Preface

Over the past few decades, exciting developments have taken place in the field of near-infrared spectroscopy (NIRS). NIRS has grown from an initial small niche market to become an indispensable sensor. This has been enabled by the advent of robust Fourier transform interferometers and diode array solutions, coupled with complex chemometric methods that can easily be executed using modern microprocessors. Nowadays, NIRS has proven to be a reliable and an inexpensive method with a great potential for use in the process industry, for advanced control and product quality assurance, as well as for medical applications with particular focus on diagnostics.

The present edited volume intends to cover recent developments in NIRS and provide a broad perspective of some of the challenges that characterize the field. The target audience for this book includes engineers, practitioners, and researchers involved in NIRS system design and its use for different applications. We believe that they will greatly benefit from the timely and accurate information provided in this work. The volume comprises six chapters overall.

NIRS possesses unique advantages as a functional imaging method particularly for brain imaging where the brain-device interface design is very important. There are several obstacles that still prevent this technology from becoming a prominent medical imaging tool. Different potential clinical applications of NIRS imaging combined with intravascular ultrasound also exist. The technology has the potential to become a valuable tool for coronary plaque characterization and predictor of future coronary events in coronary artery disease patients.

Within the agriculture field, NIRS is particularly effective in the measurement of mineral nutrients, organic compounds (including carbohydrates), and other physical and chemical characteristics for a wide range of different types of samples. This is enabled by advanced chemometrics based on complex signal filtering as well as linear and nonlinear regression methods. The detection of trace metals in environmental matrices is another particularly important application. This includes metals such as cadmium, copper, lead, chromium, and mercury, which are major environmental pollutants. Looking into the future, further research focuses not only on chemometrics and sophisticated interferometers but also on the fine-tuning of photodiode systems for bespoke system development.

The editor is indebted to all the colleagues from across the world that contributed to this volume with their latest research, to Jan Skvaril for joining this effort as a coeditor, to several NIRS and chemometrics experts who volunteered as reviewers, and to InTech for the opportunity to work on this volume and its members of staff for their constant support during its preparation.

Prof. Konstantinos G. Kyprianidis
Future Energy Center
Department of Energy, Building and Environment
Mälardalen University, Västerås, Sweden

Chapter 1

The NIRS Cap: Key Part of Emerging Wearable Brain-Device Interfaces

Amal Kassab and Mohamad Sawan

Additional information is available at the end of the chapter

http://dx.doi.org/10.5772/67457

Abstract

Nowadays, near-infrared spectroscopy (NIRS) fills a niche in medical imaging due to various reasons including non-invasiveness and portability. The special characteristics of NIRS imaging make it suitable to handle topics that were only approachable using electroencephalography (EEG) such as imaging infants and children; or studying the human brain activity during actions, like walking and drawing that require a certain amount of freedom that non-portable devices such as magnetic resonance imaging (MRI) cannot permit. This chapter discusses the unique advantages of NIRS as a functional imaging method and the main obstacles that still prevent this technology from becoming a prominent medical imaging tool. In particular, in this chapter we focus on the design of the brain-device interface: the NIRS cap and its important role in the imaging process.

Keywords: NIRS cap design, fNIRS, NIRI, medical imaging accessories, portable brain imaging, optode holder

1. Introduction

Near infrared spectroscopy (NIRS) has been gaining momentum due to its unique advantages that makes it an indispensable tool in medical research. By successfully resolving certain issues of portability and data filtration, NIRS is expected to find a wide application not only in medicine but also in the gaming industry as well as any thought controlled electronic devices due to its relatively inexpensive, portable and non-invasive nature. From a medical standpoint, the advantages of NIRS imaging, or functional NIRS (fNIRS), are quite distinct. Much like electroencephalography (EEG), its portability and non-invasiveness make it a natural choice for imaging young children and infants [1]; however, while EEG signals are inherently noisy, non-linear and rely on electrical signal variations on the scalp [2], NIRS offers 1–2 cm depth resolution that is capable of capturing cortical activation [3]. Additionally, NIRS offers higher temporal resolution than traditional immobile imaging devices such as functional magnetic resonance imaging (fMRI) and positron emission tomography (PET), which allows the detection of transient cortical events [4]. Undoubtedly, present imaging techniques in general are bound to offer higher temporal and spatial resolution as their design develops over time, but what makes NIRS imaging an interesting contender is the combination of the previously mentioned factors which allows it to be a suitable device for long-term cortical activity monitoring. NIRS promises a device that can be used anywhere, inside or outside of a lab or hospital setting and that can register cortical activation throughout different activities with varying degrees of freedom without particular concern towards the subject's age group or physical condition which can have important real-life applications today [5–8]. Nevertheless, for NIRS to achieve its full potential the topic of its interface is yet to be properly addressed and designed.

The application of NIRS imaging relies on two primary factors: the first factor is the relative transparency of human tissue to near infra-red (NIR) light, which penetrates the skin, subcutaneous fat, skull and brain [9]. The second factor is the high attenuation of NIR light due to haemoglobin oxygenation levels [3]. More specifically, the term 'optical window' is used to define the range between 650 and 1350 nm where light absorption coefficients of water, melanin in addition to oxy and deoxy haemoglobin, are lowest. This allows a certain amount of light to penetrate biological tissue, where it is scattered and eventually diffused allowing for a limited amount of tissue penetration to occur. NIR imaging relies on light absorption coefficient values of key biological components, such as water, oxy and deoxy haemoglobin to measure changes in their concentration over time. For example, as shown in **Figure 1**, the absorption coefficients of oxy and deoxy haemoglobin intersect at around 805 nm allowing for the use of two distinct NIR wavelengths within the optical window to measure the changes in each of these elements [10–12].

Pigmented compounds such as chromophores of skin and hair melanin are also a high source of NIR attenuation; however, these factors are easily corrected by adjusting light intensity since their value over the period of imaging is constant [13]. The behaviour of NIR light inside tissue is also relevant, as the main mechanism of NIR light propagation is scattering, and while a part of NIR light is attenuated as it is absorbed by chromophores, the remaining scattered photons resurface back a certain distance away from the light source allowing the detection and measurement of light attenuation over time.

Since NIRS allows the measurement of oxy and deoxy haemoglobin changes over time, it is considered an indirect method of measuring brain activity based on the neurovascular coupling phenomenon that relates neural activation with vascular response. Neurovascular coupling refers to the increase in oxy-haemoglobin (HbO) and simultaneous decrease in deoxy-haemoglobin (HbR) when spatially clustered 'cortical columns' that share the same functional properties are stimulated. This cluster formation is what makes brain oxygenation levels detectable using optical imaging [14, 15].

NIRS can also provide non-haemoglobin-based measurements, by recording data from several wavelengths simultaneously, in order to detect tissue chromophores, including cytochrome oxidase the marker of metabolic demands [16]. While some studies suggest the use of

http://dx.doi.org/10.5772/67457

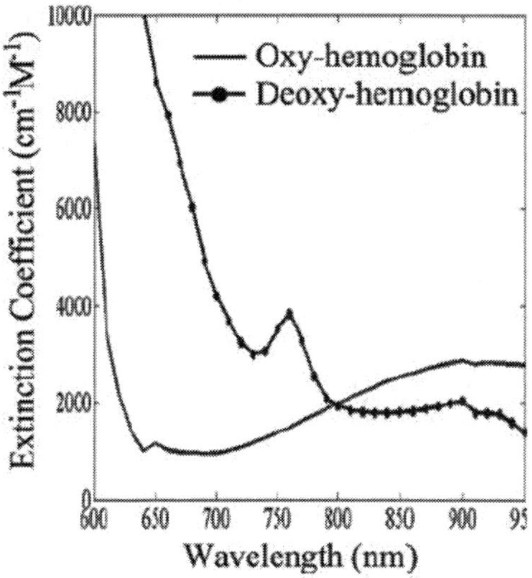

Figure 1. Light absorption spectrum of oxy and deoxy-haemoglobin, the span between 650 and 950 nm is called the 'optical window' due to the relatively low absorption factors in tissue [3].

NIRS in the detection of cell swelling as a result of neuronal firing in order to directly detect neuronal activity; however, these signals are 0.01% smaller than hemodynamic activity making it a less reliable method for detection [17–19].

Overall, although the special characteristics of NIR light were first published by Jobsis in 1977 [9], yet the first 10-channel NIRS imaging system was only introduced in 1995 and actual interest in this technique was only seriously considered with the advent of multi-channel wearable and wireless devices in 2009 [20]; since then NIRS has been used extensively in brain imaging research which is reflected in the number of publications that cover its development, use and various applications today. Nevertheless, NIRS has low reliability still in single subjects, which makes it unsuitable for clinical applications and restricts its use in large group medical research [21–24].

2. fNIRS instrumentation

Any NIRS device can be divided into three major components: (1) a brain device interface that includes optodes and the cap stabilizing them, (2) a control module that collects, sorts registered data and provides the various illumination schemes in addition to data transfer to the (3) user interface and main software responsible for analysing data using signal processing algorithms.

These three branches will be discussed briefly; however, emphasis will be on the interface and the essential role it plays on the imaging process.

2.1. The brain-device interface

The term 'optode' refers to both the NIR light emitter and detector that ideally create a fixed and predetermined illumination scheme within the cerebral cortex. The source, or light emitter, shines light directly into the scalp, this light is scattered by head tissue causing it to deflect in all directions and only a small fraction of this light (approximately one out of 10^9 photons) resurfaces back to the scalp some distance away from the entry position [3]. This NIR light distribution was simulated by Okada and Delpy, their study showed the light-scattering pattern within the scalp, skull and cerebrospinal fluid in addition to the sensitivity of each source-detector pair to this scattering, which creates a banana-like shape within the scalp with two narrow ends at the source-detector locations [25]. Light attenuation can be calculated based on the Beer-Lambert law that links the ratio of incident and reflected intensities to the absorption and diffusion phenomenon [9]. On the other hand, the distance where the NIR light resurfaces back differs from one subject to another based on age, curvature of the scalp and head size and it generally ranges from 3 to 4 cm; therefore, an ideally placed light detector at that exit position can capture it. The change in the amount of detected light overtime is used as an indicator of the absorption variation of NIR light due to cortical activation.

Based on this light emitter-detector coupling, also called 'channel', it is clear that unlike other non-invasive brain imaging techniques, such as EEG, the integrity of a NIRS signal relies on the assumption that the cortical illumination and detection scheme is ideal. This assumption entails that the relative position of the detector/emitter couple is constant, and that the detector, emitter operational conditions are constant throughout the imaging session. However, this is often not the case, and so far, it has been a very difficult condition to maintain, particularly for the type of experimental requirements that fNIRS is designed to meet such as imaging freely moving subjects over extended periods of time.

Most successful fNIRS imaging experiments are commonly conducted inside a lab, where the subject sits still on a chair and is refrained from talking, smiling or moving their head. With the advent of better signal filtration, successful use of fNIRS was also registered in rehabilitation centres with walking patients, or even cycling [5, 26, 27]; however, constrains on facial expression and subtle head movements still apply, because while certain movement artefacts such walking and running and obvious head movements are easier to isolate and/or filter out, facial expressions are far more difficult to detect. Small facial muscular fluctuations or hair resistance to NIRS optodes that are unnoticeable to outside observers can cause the entire optode holder to slide or cause slight optode inclinations. Such inclinations that fluctuate over the imaging period can cause light scattering outside the scalp, poor light detection or displacement of surrounding hair in front of optodes resulting in false attenuation values that cannot be accounted for using common artefact detection methods. Therefore, subjects with dark and voluminous hair are typically the hardest to image as any displacement of hair in front of the optodes can jeopardize the integrity of the results while voluminous hair adds resistance and counter pressure against the optode holders.

http://dx.doi.org/10.5772/67457

As mentioned previously, optical penetration is usually 1–2 cm, which is typically half the source-detector distance. Such penetration depth translates to 5–10 mm of outer brain tissue penetration after subtracting the thickness of the skin, subcutaneous fat and skull (which vary from one person to another and with age), this allows the detection of the outermost cortex activation [3, 10]. Most NIRS devices rely on two light wavelengths simultaneously to measure both oxy and deoxy haemoglobin changes [28–30], while three or four wavelengths might be used in some cases in order to either extract changes in other species, such as water and lipids [31, 32], or to couple with time resolved methods for additional parameters such as blood flow and absolute tissue saturation [19].

There are various types of NIR light sources, the two most commonly used emitters today are laser diodes and light-emitting diodes (LEDs). Laser diodes provide a technical advantage over LEDs as they have higher light intensity and smaller optode size, which allows for better hair penetration and scalp contact. However, they have higher energy consumption and cost, thus their use is not suitable for portable devices outside a lab environment. LEDs require simpler circuitry; they generate a light spectrum of about 30 nm and are the natural choice so far for portable fNIRS systems [19].

As for light detectors, the most common choice is avalanche photodiodes (APDs) that translate the amount of detected photons into current and have low power consumption with the capacity to increase the detected light intensity. In addition, APDs are fast with more than 100 MHz speed and have a high sensitivity with the dynamic range of approximately 60 dB. Some devices rely on silicon photodiodes, however, these have a medium speed and lower sensitivity but a higher dynamic range with approximately 100 dB [19]. Modern microfabrication techniques are aiming at the creation of smaller LED and APD designs with enhanced capability, which is essential to the development of next generation portable fNIRS devices.

Finally, when it comes to optode holders, there are two major types of optode stabilizing methods, the cap (a soft headwear) that covers the entire head, with prefixed locations for optodes, much like an EEG cap. However, in NIRS caps, the optodes are not prefixed on the cap in order to allow for hair manipulation and tossing to take place prior to optodes installation. The other common types of optode holders are the rigid patches that cover a certain cranial zone. The term 'rigid' refers to the material used for stabilizing the optodes, since although they are made of silicon which allows it to bend slightly to fit the head shape at a given location, the distance between the optodes is fixed as the material itself does not stretch, unlike the cap, thus the distance between the optodes is fixed throughout the imaging session giving the patches a clear advantage over the caps. Both designs are prone for sliding, however, requiring additional restrains to keep them in place, such as attachments under the chin or to a belt that goes under the armpits and over the chest.

2.2. The electronic control module

The electronic control volume is directly connected to the optodes and therefore is the portable part of the fNIRS device in addition to the interface. This component is responsible for the illumination scheme in addition to data gathering and transmitting (in portable devices).

Lighting strategies in fNIRS aim to reduce power consumption and heating of the scalp in addition to differentiating between various light emitting sources, which is essential to distinguish between the different channels when there are multiple light emitters within the range of a single detector. Therefore, the control module employs a certain method for multiplexing and/or modulating of light sources.

However, the most significant aspect of the control module is its illumination technique. There are three major types of illumination schemes used today which are shown in **Figure 2**. The most common type is continuous wave (CW) which measures simply the backscattered light intensity attenuation. The second type is the frequency domain (FD), which uses intensity-modulated light in order to measure both attenuation and phase delay of returning light. The third technique is the time domain (TD), which relies on short pulses of light as an illumination source and detects the shape of the pulse after propagation through the tissue; this technique provides information about spatial specificity in addition to tissue absorption and scattering [33].

The CW scheme is relatively simple and cost effective as it relies on establishing a baseline, or a zero state, and then compares oxy and deoxy absorption changes to this initial value during a certain test or a task. However, only FD and TD methods can provide absolute characterization of tissue properties including the distinction between absorption and scattering in the tissue [20]. Nevertheless, a more complex scheme is generally associated with lower time resolution and is more susceptible to noise and movement artefacts, since determining the

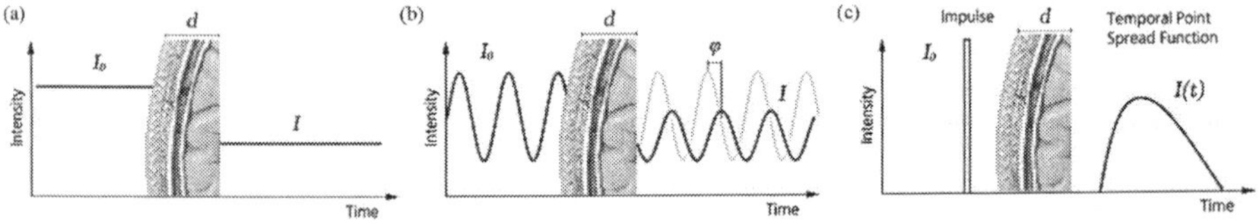

Figure 2. The three type of fNIRS illumination techniques: (a) continuous wave, (b) frequency domain and (c) time domain (TD).

Main characteristics	Continuous wave	Frequency domain	Time domain
Sampling rate (Hz)	≤100	≤50	≤10
Discrimination between cerebral and extra-cerebral tissue	Not possible	Feasible	Feasible
Measuring HbR, HbO	Only changes	Absolute value	Absolute value
Measuring scattering, absorption coefficient and pathlength	No	Yes	Yes
Measuring tissue HbO saturation (%)	No	Yes	Yes

Table 1. The main differences between the three different illumination types in fNIRS (adopted from Ref. [20]).

time of flight is effected with geometrical and contact changes. The major differences between the three techniques are summarized in **Table 1**.

It is important to keep in mind that as fNIRS fills a special niche for portable imaging systems, the most important qualifications in general are those related to power consumption and size, which explains why most fNIRS devices adopt the simplest illumination technique. In addition, present application of fNIRS does not require tissue characterization as it is more concerned with changes in blood oxygenation rather than absolute absorption values [3]. However, both of these aspects might change as fNIRS reliability is increased and the technology used in FD and TD systems becomes more compact and power efficient.

2.3. Data analysis and user interface

This is where data from each illumination channel are gathered in order to be filtered, quantified and presented in a user friendly fashion. It is also where certain controls over the system in general are provided from the end user as actions and input variables. The fNIRS software package is usually provided on a computer or even a tablet with a Bluetooth connection to the control module.

There are many algorithms and software dedicated to optical imaging and signal quantification based on how light behaves in tissue. The two most widely used theoretical models are the differential pathlength factor (DPF) and the diffusion approximation. Both assume that tissue is homogeneous, however, the diffusion approximation method assumes that scattering is larger than absorption; therefore, each type of tissue has a specific geometry (infinite, semi-infinite, slab or two-layered) [34, 35]. Still, since the two models rely on quantification over a given path, interpersonal differences such as the thickness of scalp, skull and cerebral spinal fluid in addition to hair and skin melanin concentrations are bound to create biases in spatial localization of brain activity particularly with TD and FD methods, but they are less significant in CW methods [25, 36, 37].

This chapter will not cover all the various aspects related to the proper functionality of this component, it will only concentrate on aspects related to noise attenuation and filtration for their obvious relation with the signal quality obtained that is provided by the device interface and is affected by the cap design.

In general, noise sources can be either instrumental, experimental or physiological. Instrumental and experimental artefacts refer to experimental errors including movement artefacts and device malfunction and have to be dealt with prior to data analysis. Physiological errors on the other hand are due to certain changes in the physiology of the subject that affect but are not part of the intended experiment. These are usually treated with filters after the conversion of raw signals to haemoglobin units either using algorithms that compensate for pulse-related artefacts or by using additional NIRS channels that measure extra-cortical hemodynamic variations [3, 10–12, 38]. Instrumental errors have to be dealt with prior to any testing, since they can easily overpower the measured signals. Whereas movement artefacts should be approached by carefully controlling the experimental environment whenever possible. However, since absolute control over the entirety of the experiment is not likely, not to mention that the nature of the experiment itself might produce movement artefact, such as walking or cycling, special algorithms have been developed to filter out these errors using additional data collecting methods, such as a camera [39] or an accelerometer [38, 40].

Nevertheless, to date there are no methods that can provide any information regarding optode-scalp contact 'quality' to ensure that the received data reflect that of an ideal illumination condition throughout the imaging session.

3. fNIRS caps: objectives and challenges

Clearly, the primary objective of the NIRS cap is to stabilize the optodes, making sure that they are in constant contact with the scalp throughout the imaging period. However, in practice, there are other concerns that affect the proper functionality of the NIRS cap and its future use, namely: the installation process and subject comfort.

The effect of optode stability on fNIRS signal quality was not quantified until recently, when the work presented by Yücel et al. was published in 2013 and 2014 [41, 42]. In these studies, the authors glued fibre optic optodes on the scalp using collodion, which is normally employed with EEG electrodes to monitor epilepsy patients. Using this method, the authors reported 90% reduction in signal change due to movement artefacts, a signal-to-noise (SNR) increase by sixfold and threefold at 690 and 830 nm wavelengths, respectively, and a statistically lower change in both oxy and deoxy haemoglobin during movement artefacts. In spite of the fact that their optode stabilizing methodology may not be practical for short-term and off-hospital settings. Nevertheless, this study provides an objective assessment of the effect of interface stability on the fNIRS signal, especially with moving subjects.

Nevertheless, the task of stabilizing the optode using a mechanical device is quite elusive due to several reasons:

1. Current optode stabilization techniques rely on pressure; however, pressure is also a major source of discomfort. Thus, the more stable the optode, the more discomfort it is bound to create for patients. Such conditions might be tolerable for short-term monitoring periods of 10–20 minutes; however, as the imaging session becomes longer these stabilizing techniques may not be acceptable. Presently, there are no studies identifying the comfort pressure threshold on the scalp, although such studies were done for other anatomical parts of the body [43]. Additionally, pressure values necessary to stabilize the optode are also unspecified yet. Preliminary results indicate that comfort pressure values on the scalp are not uniform, as the forehead and the back of the head, particularly the area behind the ears tend to be more sensitive than other areas on the scalp. More importantly, the difference between the pressure needed to provide optode stability (approximately 30–45 Pa) versus the comfort pressure margins on the head (50–60 Pa) is very small [44], therefore, designing an optode holder that relies solely on pressure is quite a challenging task. It is important to mention at this stage that both the comfort pressure as well as the pressure values necessary to stabilize the optode are tentative preliminary results and that such claims can only be established once a study on a large number of participants is conducted. In general, the results obtained from the preliminary study are in accordance with lab observations as fNIRS results tend to be better with less comfortable and higher pressure inducing headwear.

2. Since the importance of having a tight headwear at all cranial positions has been well clarified, one of the major obstacles in designing an ideal fNIRS cap is presented in head shape variations from one subject to another. Such variations can even be present within the same subject as differences between the right and left side of the cranium might exist. These often cause uncomfortable high pressure areas versus 'pressure gaps' where the optode fails to provide the necessary force to maintain scalp contact or prevent surrounding hair from covering the optode. While imaging companies try to compensate for general head shape variations by providing three (or more) headwear sizes (small, medium and large); even introducing different designs for certain markets in order to compensate for head shape differences between several ethnicities [45]. However, interpersonal head shape variations cannot be accounted for and simple caps often cannot meet the basic requirement of providing a perfect fit for all subjects. Partial head covering patches may present a reasonable solution in cases where imaging the entire head is not required, as their size allows for a certain degree of manipulation over the required imaged zone. However, such patches are prone to slippage and require extra attachments to keep them in place.

3. The third factor in assessing an optimal cap design is the difficulty associated with its installation. While fNIRS cap installation is considered a cumbersome task that necessitates an expert technician, it is important to keep in mind the anticipated goals of a portable brain imaging system, including its role as a brain-device interface with applications spanning from gaming to medical devices. Therefore, unassisted single person installation is the ultimate goal for future fNIRS applications, albeit it is far from becoming reality with present designs.

Today, the installation of the fNIRS cap can be a long process that starts with taking general head measurements to identify important reference locations based on the 10/20 system. This is followed by the placement of the patch or cap and documenting the distance of the optodes from this (these) reference points, then rigorous clearing of hair at various optode locations is performed, and finally the optodes are placed. This process may take up to one hour based on the cranial area covered and the type as well as the amount of hair present. Therefore, attempts at creating easier installation of optode holders invariably address easier hair tossing or clearing methods, since this is generally the most time-consuming part of the process. Apart from providing certain clearances around the imaged zoned to easily toss the hair (particularly when using the patches) the only solution so far seems to be in creating smaller optodes that would infiltrate hair to ensure optimal scalp contact in addition to increasing localized optode pressure by an in-house spring. These solutions assume that optode size can eventually decrease to a point where it can become comparable to hair strands. However, this is far from the actual optode design available presently.

Based on these observations, it is clear that traditional fNIRS caps cannot meet the demanding requirements of portable fNIRS-based imaging. But before proceeding to possible future solutions, the next section will focus on fNIRS cap designs that were developed so far in the literature and whether possible solutions can be based on these proposals.

4. The design of the optode holder: an overview

The design of the fNIRS cap has not received much attention in research or in the literature. This was due mostly to the fact that the fNIRS device is purely an electronic one, thus it elicited the focus of electrical and optical engineers and physicians while the optode stabilizing method itself, a mechanical device, was mostly dealt with as an accessory. The NIRS cap and the installation process were mentioned in 2009 by Huppert et al. for the first time, where the author voiced the importance of stabilizing the optodes and its effect on reducing experimental errors. The authors suggested more anchoring methods to attach the head band to the body in order to reduce the effect of the weight of the optodes on motion instability. They were also the first to mention the important dilemma of subject comfort during imaging due to the additional restrains [3].

The design proposed by Huppert et al. is shown in **Figure 3**, and it portrays the stretchable cap that is used to stabilize a polymer patch which acts as an optode holder. Thus, the cap provides both a rigid spacing for the optodes and a flexible material to hold the patch in place, with additional Velcro attachments to stabilize the optodes and their wiring. The authors specify that even more rigorous attachments are needed for moving subjects. The design was made for in-lab fNIRS measurements; therefore, the stability it provided with moving subject was not demonstrated.

Apart from this example, other attempts to create a head band for the prefrontal area were also introduced in 2009, where no complications due to hair interference can be found and the stability of the head band can be controlled by simply increasing the amount of pressure by changing the size of the head band. One such design is presented by Atsumori et al. [46]. While similar designs may be useful for gaming applications, in addition to few medical and research studies that focus on the prefrontal cortex, however, the bulky

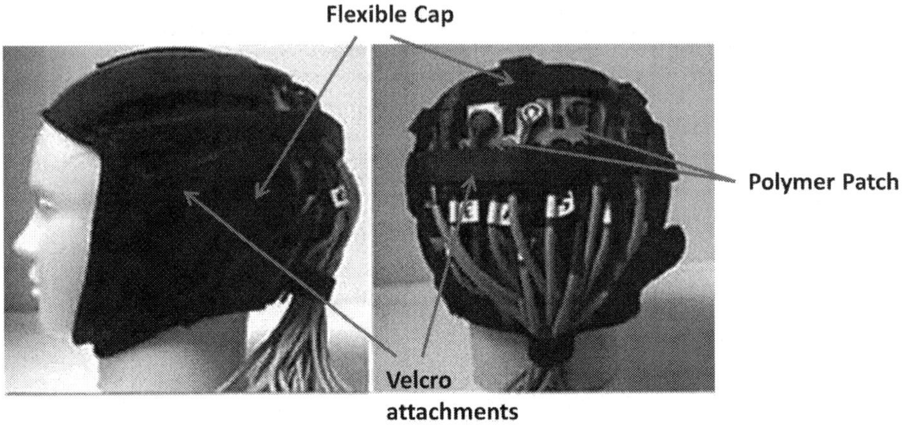

Figure 3. Stretchable cap design that holds a flexible polymer patch and stabilizes it with Velcro attachments [3].

http://dx.doi.org/10.5772/67457

design represents additional mass that would contribute to movement artefacts, also the fact that it relies solely on pressure to ensure stability makes its use restricted to short-term applications.

Another study for an fNIRS cap was presented by Kiguchi et al. in 2012 [47], using a cap that was made of a black rubber. This might be considered the earliest study dedicated to the fNIRS cap for 'haired' regions including the design of the optodes. The optodes in the helmet like cap are fixed on the inside, as an integral part of the helmet that cannot be accessed or manipulated by the end user. Instead the authors chose to stabilize the optodes surrounding hair by rubber teeth. These teeth aim also to reduce the discomfort presented by optode localized pressure that was induced by a spring. Although this study is dedicated for portable fNIRS devices, however, it does not mention neither the installation process nor present a comparative demonstration of the stability it provided to the optodes versus other cap designs. Nevertheless, the bulky design does not present a practical solution against weight-induced movement artefacts, additionally, holding the hair in place does not guard against slippage or blocking the NIR light by hair in front of the optode.

Regardless of the success of the design proposed by Kiguchi et al. the idea of using a glass rod to reduce optode-scalp contact area which results in less optode resistance by surrounding hair has been adopted in the first open-air fNIRS study published by Piper et al. in 2014. This study also provides the first comparative look at the effect of movement artefacts on signal integrity. The imaging quality was tested under three different conditions that varied between indoor sitting on a stationary bike, indoor pedalling on a stationary bike and outdoor bicycle riding [26]. The fNIRS cap used in this study is the regular EEG-inspired elastic cap that has been available in the market for sometime. However, innovation lies within the minimization of optode size that is further reduced by the use of a 3 mm in diameter glass rod to guide the light into and from the scalp, in addition to reducing the weight of the connecting optode wires. Although this improved design has allowed the implementation of fNIRS imaging outdoors, still movement artefacts affected the fNIRS signal visibly as demonstrated by the study. As rejected channels per person were only 5% for someone sitting on a stationary bike, but this value increases to 7.5% during indoor pedalling and reaches 35% for outdoor cycling [26]. Obviously, a different approach for designing optode holders is warranted.

A comprehensive study on the design of an optimal fNIRS cap was provided by the work of the Imaginc group, in order to explore several design ideas that targets the issue of patient comfort and signal stability [11, 12, 44, 48]. Their study showcased several concepts ranging from padded fNIRS caps/helmets that were geared towards patient comfort, to Velcro patches that provide a none flexible alternative to stretchable caps with an option of adding strands or adjusting for size based on the subject's head shape; in addition to stretchable elastic bands that provide extra space for hair tossing and ventilation, as shown in **Figure 4**. The study concluded that designs that focused on patient comfort as a primary goal failed completely in providing the necessary grip for optode stability. While designs that focused on optode stabilizing based on applied pressure were relatively successful and their success was a function of the amount of pressure it provided on the participant's scalp.

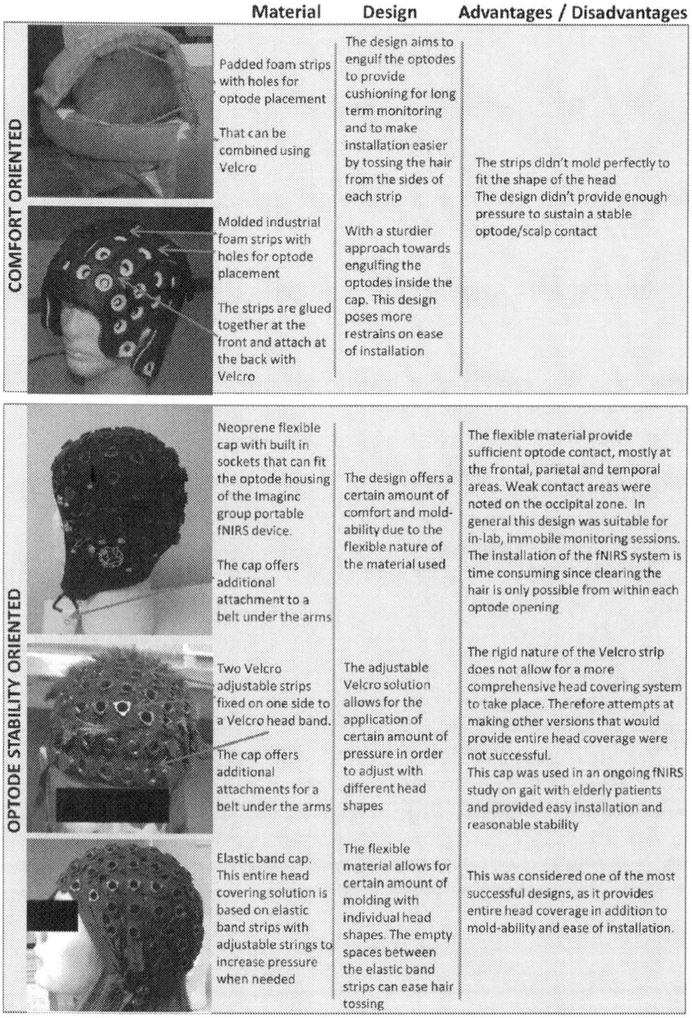

Figure 4. Different headwear designs for optode holders, comfortable versus stable cap designs.

The direct correlation between pressure and signal stability regardless of cap design was clearly demonstrated in a comparative experiment between different cap models that were developed by the Imaginc group. The most successful models that were tested included the Velcro cap, the elastic band cap and the neoprene cap. Movement artefacts were recorded while the subject was sitting motionless, as a baseline, then while moving the head backwards, forwards and sideways followed by a period of walking. The results obtained are shown in **Figure 5**, that also note the number of rejected channels in each case. Surprisingly, in spite of previous results that have restrained the use of the neoprene cap to stationary in-lab testing, while the Velcro and elastic band caps were more successful with freely moving subjects, the neoprene cap presented surprising noise artefact reduction, even while moving the head. This was due to the fact that the cap was too tight and visibly uncomfortable for the user, which is a clear indication of the inverse relation between optode stability and comfort. On the other hand, the effect of head movement on motion artefacts was much larger than walking, even without using motion filtration methods.

This led the team to explore other methods to stabilize the optodes that do not rely entirely on localized pressure. These proposals will be reviewed in the following section.

Number of Rejected Channels in each case

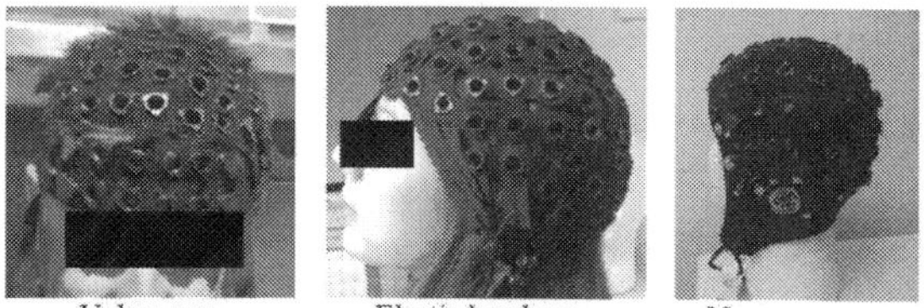

	Velcro cap	Elastic band cap	Neoprene cap
Sitting	69	130	140
Head movement	74	138	146
Walking	73	135	143

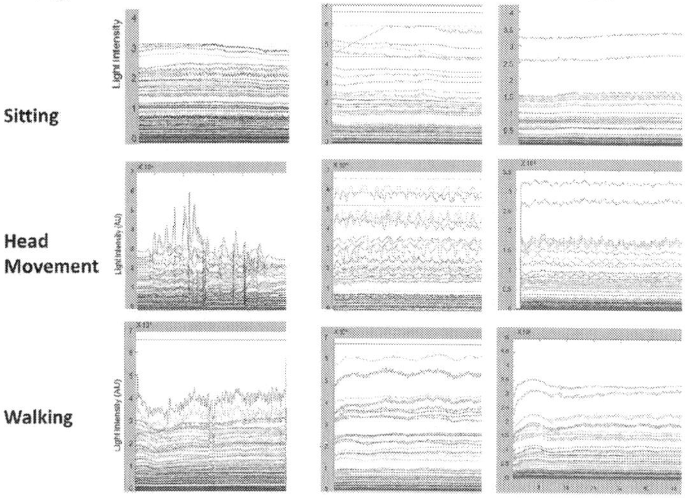

Sample Reading for Each Cap Design

Figure 5. Comparative look at the various cap designs and motion artefacts under different conditions: sitting, head movements and walking.

5. Gripping the head: the future of fNIRS caps

Previous studies conducted in the field of fNIRS cap design lead to an obvious conclusion, relying on pressure alone as a means of stabilizing the optode is not a good strategy when it comes to imaging applications. Conversely, the science of providing a perfect grip for any object is not a new one particularly in the field of robotics. Indeed, robotic arms that are being developed for several applications ranging from the industrial to the medical have already crossed several milestones in achieving gripping capabilities against slippage in addition to handling sensitive objects with speed and accuracy. In reviewing the vast literature published in this field, it is possible to find a couple of comparable solutions that can provide the required amount of grip, mould-ability with individual head shapes and ensuring patient comfort at the same time [48].

While handling sensitive objects, a firm grip is often associated with engulfing the gripped item in order to create a distributed pressure force instead of localized ones; additionally, engulfing the gripped object creates friction, which is the horizontal force that prevents slippage [47]. The more surface area of the object is covered the better hold the gripper provides with improved protection against slippage. Technologies such as the 'universal gripper', for example, can firmly hold a raw egg without breaking it. While on the other end, sensor-equipped artificial hands provide a perfect grip using accurate feedback of the amount of force applied on each point. The difference between these two technologies is vast, as the universal gripper is an extremely simple solution that relies on moulding the gripper to fit around the object, by the use of a simple grain-filled elastic bag and a vacuum pump [49]. On the other hand, the sensor-equipped gripper requires numerous actuators, a processor and sensors to perform properly [50].

Since the quality of the grip is as important as the force that is required to provide it. The two previous methods can translate into pneumatic solutions that are promising for fNIRS imaging. The sensor-based system although costly offers an important additional feedback input that has been thus far lacking in present fNIRS systems, the quality of contact: or in other words, the amount of pressure at each optode location. Thus, defining the optimal optode pressure becomes an important factor in such systems and can help filter out signals when optode pressure values are below a certain threshold. Such a system can have the inflatable cap structure proposed in **Figure 6**. The two-layered air tight cap should be made of two different polymer types, with a more elastic one at the interior in order to allow for maximum moulding and expansion on the inside of the cap rather than the outside. Additionally, the interior of the cap should be lined with pressure sensors that provide feedback to a microcontroller. Based on the return signal, the microcontroller changes the state of the valves (either open or close) in order to inflate the air pockets. It also controls the air pump that inflates the balloons and turns it off once all the valves are closed. Dividing the cap into several air pockets is also an important part of the design, since interpersonal variations in certain cranial zones are less than in other areas.

The inflatable pockets can help also provide optode cushioning, which not only increases its stability but also allows for overnight use of the fNIRS imaging system.

http://dx.doi.org/10.5772/67457

However, this solution presents technological challenges, such as the fabrication of sensors, micro-valves, miniaturized pump and controller. In addition to ensuring that the power consumption of the added electronic components is minimal and that the cap's weight is low.

The second pneumatic solution, on the other hand, requires a vacuum pump that is not necessarily integrated in the cap itself, and it can be considerably less challenging from a technological point of view. One example of a vacuum fNIRS cap that is an adaptation from the universal gripper concept is presented in **Figure 7**. As shown, the cap can be a regular fNIRS headwear that is lined with small grain-filled balloons, or it can be made of an air tight polymer that covers the entire head also filled with small grains (the examples shown are filled with coffee or small foam grains). No embedded electronic components are necessary or required for this solution, instead, the cap can be firmly placed on the participant's head, then

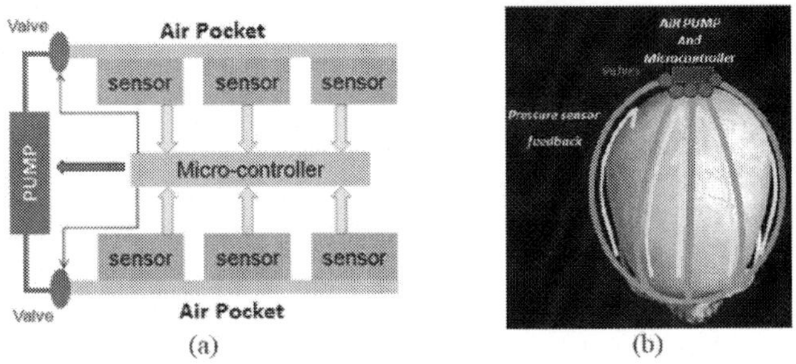

Figure 6. Inflatable pneumatic cap design: a) a schematic representation of the various inflatable cap components b) a top view showing the location of the various components on the head. The cap is divided to several air pockets that are lined up with pressure sensors, once the return signal from all pressure sensors at a given air pocket are above a certain value, the microcontroller closes the valve of that air pocket, and once all air pockets valves are closed the pump is turned off too [44].

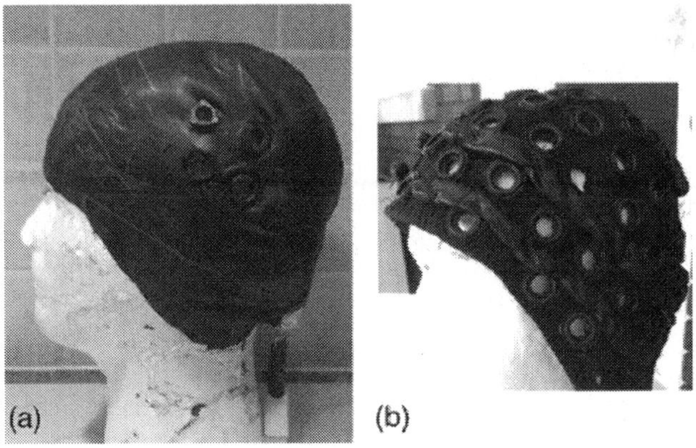

Figure 7. Vacuum fNIRS cap design (a) an airtight latex cap covering the entire head and filled with granular foam balls, (b) a regular fNIRS cap lined with tube balloons filled with coffee grains [44].

the vacuum pump is used to 'mould' the cap and jam the grainy material in order to create a tight grip on the head. Once this is achieved, hair can be tossed securely and the optodes can be placed in their designated sockets. The cap firmly holds the head until the imaging session is over then air is allowed back into the cap thus loosening its grip.

With such gripping methods, the cap is expected to be worn without the need for additional attachments connecting the cap to a belt under the arms or to the chin. However, without an actual demonstration of the stability of these designs, such expectations remain speculative. Preliminary results from the vacuum cap design indicate that the complete head wear provides a tighter grip than the balloon-lined design due to the increase in the gripped surface area.

So far, the topic of future fNIRS caps focused on user's comfort and optode stability. However, cap installation is an important part of anticipated fNIRS imaging applications, not to mention its present day relevance for medical research considering the time and effort it requires from experts in the field of imaging. This has been generally due to the assumption that once smaller optodes were designed, the need for hair clearing would diminish; therefore, no mechanical methods would be necessary to clear the hair and ensure optode/scalp contact. Additionally, as optode holders generally provide a very small space (0.7–1.2 mm in diameter) to place the optodes, this complicates the design of a mechanical hair tossing device to operate in. Consequently, designing a hair clearing optode holder can be a very expensive endeavour. So far, the best caps or optode holders from installation point of view were considered the ones that provided a clearing around the optode location to help with the hair tossing process. Therefore, small polymer patches or the elastic band cap in addition to the adjustable Velcro strips cap in **Figure 4** are preferable to other complete head covering models. Still, the installation process even with the help of adjustable patches or additional spaces around the optode requires the help of an expert technician, as the only advantage they provide thus far is that of reduced installation time.

Although the topic of the importance of a hair clearing optode holder is debatable given its complexity, clearing and holding the hair in place can potentially help in stabilizing the optode holder itself. Therefore, the basic concepts for such a device will be mentioned here for future references. As shown in **Figure 8**, hair tossing can be performed using either a double hair tossing pins, a single hair tossing pin or multiple pins directing hair from the middle of the opening outwards. Such mechanisms can be added to the socket, which is the locking mechanism used to place the optode on the cap. Integrating a hair clearing mechanism that can be activated by simply placing the optode inside the opening can potentially allow for single user installation, without the need for an expert technician.

Applying these clearing techniques on hair is faced with certain complications, such as hair directionality; therefore, for a socket that has two pins, parting the hair from the middle cannot be helpful at locations where hair direction is not parallel to the pins. This is even more complicated with one pin parting designs, therefore 'adjusting' methods for

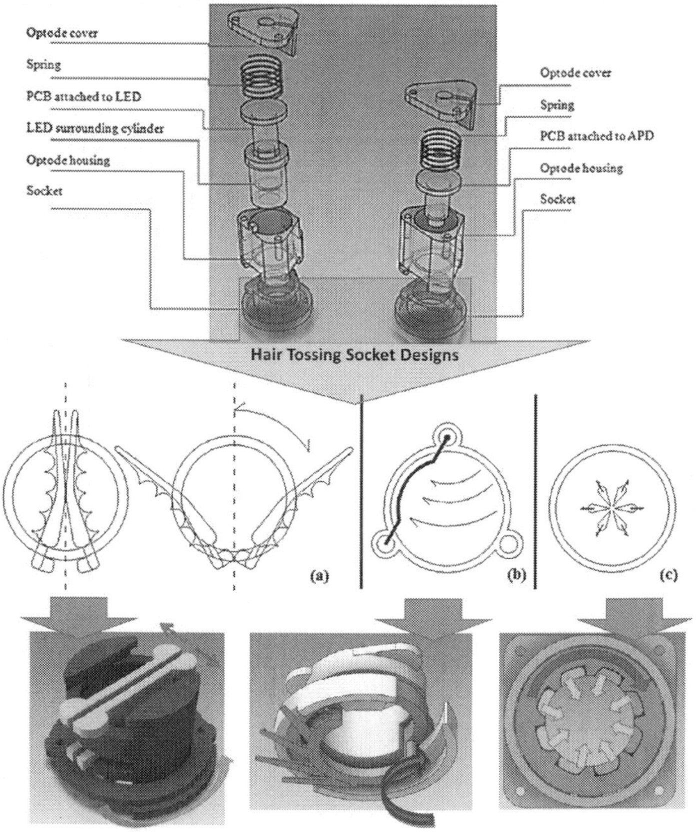

Figure 8. The various components of the optode housing and how it connects to the socket that is attached to the cap, a spring located inside the optode housing provides an additional pressure to maintain optode/scale contact. Hair clearing can be achieved via the development of socket designs that can play a dual role, by adding a hair clearing mechanism to it. Possible hair parting methods are: (a) dual parting pins, (b) single parting pin and (c) multiple parting [48].

pin directionality are necessary, such as designing a pin that can be placed in any of three placement combinations. Ideally, parting the hair from the middle is the best method, as it eliminates all difficulties associated with the other two methods. From a practical standpoint, the size of such pins would be in the millimetre range (maximum a centimetre); therefore, any concept needs to be tested on various levels, mechanical design, machining and implantation in order for it to be viable. Preliminary results obtained from the study presented by Kassab [48] shows that a collet-based pin concept that parts the hair from the middle by a simple twist of the optode holder can provide an interesting solution for single user applications. Such a design can be very simple to use, would not require knowledge about hair directionality and is not affected by a hair type. More importantly, it can potentially lock the hair around the optode holder thus providing additional cap stabilizing mechanism.

6. Conclusion

The aim of this chapter is to demonstrate the importance of fNIRS caps or optode holders as an interface, and how the imaging signal and ergo the future use of fNIRS can be affected by its efficiency and performance. The major challenges of an efficient imaging cap were articulated as well as present available models and possible future solutions. In general, the field of fNIRS imaging has not been generous when it comes to studies aimed at the interface itself, albeit designing an ideal imaging cap can potentially be a major factor in solidifying the marketability of fNIRS imaging as an inexpensive medical device by increasing its reliability and creating a user friendly and practical system.

In preparing this study on fNIRS caps, it was obvious that several areas were in need of proper documentation, including basic definitions or guidelines, such as the pressure values for optode stability versus pressure values for patient comfort on the head. Such design parameters are important for any tight headwear and medical device designs. On the other hand, while there are numerous studies on movement artefact algorithms and how to filter out or control them, studies on optode inclination and detachment as a movement artefact associated with facial expressions or head movement, and how it affects the imaged signal is not yet approached. With long-term imaging, issues pertaining to the effect of sweat and heat on the imaged signal is also an important one, and when considering freely moving subjects, pressure fluctuation with motion, or the dynamic pressure, on the head and how it correlates with motion artefacts can also present an important feedback defining sources of error and isolating factors that have affected the reliability of fNIRS imaging for the past couple of decades.

Finally, when it comes to testing the efficiency of fNIRS cap designs, there are no protocols or standards that define its proper use and limitations. For example, some patches can be very practical with motionless subjects for finger tap testing or visual stimulation; however, they might fail with freely moving subjects. Therefore, establishing a proper testing mechanism for fNIRS caps can also aid workers and end users in understanding the limitations of each device and thus avoid possible errors in application.

Acknowledgements

The authors like to express their gratitude for the Imaginc team for their vital role in making this work possible, special thanks to Mr. Le Lan for his support and resourcefulness. The Imaginc group is supported by the Canadian Institutes of Health Research (CIHR) and the Institute of Circulatory and Respiratory Health.

http://dx.doi.org/10.5772/67457

Nomenclature

APD	Avalanche photodiodes
CW	Continuous wave
DPF	Differential path length factor
EEG	Electroencephalography
FD	Frequency domain
fNIRS	Functional near infrared spectroscopy
HbO	Oxyhaemoglobin
HbR	Deoxyhaemoglobin
LED	Light-emitting diodes
MRI	Magnetic resonance imaging
NIR	Near infrared
NIRS	Near infrared spectroscopy
PET	Positron emission tomography
SNR	Signal-to-noise ratio
TD	Time domain

Author details

Amal Kassab[1]* and Mohamad Sawan[2]

*Address all correspondence to: am_kas@encs.concordia.ca

1 Polytechnique Montreal, Montreal, Canada

2 Polystim Neurotech. Lab, Polytechnique Montreal, Montreal, Canada

References

[1] Aslin, R. N., Mehler, J. Near-infrared spectroscopy for functional studies of brain activity in human infants: promise, prospects, and challenges. Journal of Biomedical Optics. 2005;**10**(1):011009–0110093.

[2] Hoshi, Y. Functional near-infrared spectroscopy: current status and future prospects. Journal of Biomedical Optics. 2007;**12**(6):062106.

[3] Huppert, T. J., Diamond, S. G., Franceschini, M. A., Boas, D. A. HomER: a review of time-series analysis methods for near-infrared spectroscopy of the brain. Applied Optics. 2009;**48**(10):D280–D298.

[4] Lloyd-Fox, S., Blasi, A., Elwell, C. Illuminating the developing brain: the past, present and future of functional near infrared spectroscopy. Neuroscience and Biobehavioral Reviews. 2010;**34**(3):269–284.

[5] Belda-Lois, J. M., Mena-del Horno, S., Bermejo-Bosch, I., Moreno, J. C., Pons, J. L., Farina, D., Iosa, M., Molinari, M., Tamburella, F., Ramos, A., Caria, A. Rehabilitation of gait after stroke: a review towards a top-down approach. Journal of Neuroengineering and Rehabilitation. 2011;**8**(1):1.

[6] Nagaoka, T., Sakatani, K., Awano, T., Yokose, N., Hoshino, T., Murata, Y., Katayama, Y., Ishikawa, A., Eda, H. Takahashi, E., Bruley, D. F. Development of a new rehabilitation system based on a brain-computer interface using near-infrared spectroscopy. In: Oxygen Transport to Tissue XXXI. US: Springer; 2010. pp. 497–503.

[7] Watanabe, E., Nagahori, Y., Mayanagi, Y. Focus diagnosis of epilepsy using near-infrared spectroscopy. Epilepsia. 2002;**43**(s9):50–55.

[8] Zhang, B., Wang, J., Fuhlbrigge, T. Review of the commercial brain-computer interface technology from perspective of industrial robotics. In: IEEE International Conference on Automation and Logistics; IEEE. 2010. pp. 379–384.

[9] Jobsis, F. F. Noninvasive, infrared monitoring of cerebral and myocardial oxygen sufficiency and circulatory parameters. Science. 1977;**198**(4323):1264–1267.

[10] Villringer, A., Planck, J., Hock, C., Schleinkofer, L., Dirnagl, U. Near infrared spectroscopy (NIRS): a new tool to study hemodynamic changes during activation of brain function in human adults. Neuroscience Letters. 1993;**154**(1–2):101–104.

[11] Sawan, M., Salam, M. T., Le Lan, J., Kassab, A., Gélinas, S., Vannasing, P., Lesage, F., Lassonde, M., Nguyen, D. K. Wireless recording systems: from noninvasive EEG-NIRS to invasive EEG devices. IEEE Transactions on Biomedical Circuits and Systems. 2013;**7**(2):186–195.

[12] Nguyen, D. K., Tremblay, J., Pouliot, P., Vannasing, P., Florea, O., Carmant, L., Lepore, F., Sawan, M., Lesage, F., Lassonde, M. Non-invasive continuous EEG-fNIRS recording of temporal lobe seizures. Epilepsy Research. 2012;**99**(1):112–126.

[13] Ferrari, M., Mottola, L., Quaresima, V. Principles, techniques, and limitations of near infrared spectroscopy. Canadian Journal of Applied Physiology. 2004;**29**(4):463–487.

[14] Figley, C. R., Stroman, P. W. The role (s) of astrocytes and astrocyte activity in neurometabolism, neurovascular coupling, and the production of functional neuroimaging signals. European Journal of Neuroscience. 2011;**33**(4):577–588.

[15] Vanzetta, I., Grinvald, A. Coupling between neuronal activity and microcirculation: implications for functional brain imaging. HFSP Journal. 2008;**2**(2):79–98.

[16] Heekeren, H. R., Kohl, M., Obrig, H., Wenzel, R., von Pannwitz, W., Matcher, S. J., Dirnagl, U., Cooper, C. E., Villringer, A. Noninvasive assessment of changes in cytochrome-c

oxidase oxidation in human subjects during visual stimulation. Journal of Cerebral Blood Flow & Metabolism. 1999;**19**(6):592–603.

[17] Gratton, G., Fabiani, M. Shedding light on brain function: the event-related optical signal. Trends in Cognitive Sciences. 2001;**5**(8):357–363.

[18] Steinbrink, J., Kohl, M., Obrig, H., Curio, G., Syre, F., Thomas, F., Wabnitz, H., Rinneberg, H., Villringer, A. Somatosensory evoked fast optical intensity changes detected non-invasively in the adult human head. Neuroscience Letters. 2000;**291**(2):105–108.

[19] Strangman, G., Boas, D. A., Sutton, J. P. Non-invasive neuroimaging using near-infrared light. Biological Psychiatry. 2002;**52**(7):679–693.

[20] Ferrari, M., Quaresima, V. Brief review on the history of human functional near-infrared spectroscopy (fNIRS) development and fields of application. Neuroimage. 2012;**63**:921–935.

[21] Kono, T., Matsuo, K., Tsunashima, K., Kasai, K., Takizawa, R., Rogers, M. A., Yamasue, H., Yano, T., Taketani, Y., Kato, N. Multiple-time replicability of near-infrared spectroscopy recording during prefrontal activation task in healthy men. Neuroscience Research. 2007;**57**(4):504–512.

[22] Plichta, M. M., Herrmann, M. J., Baehne, C. G., Ehlis, A. C., Richter, M. M., Pauli, P., Fallgatter, A. J. Event-related functional near-infrared spectroscopy (fNIRS): are the measurements reliable? Neuroimage. 2006;**31**(1):116–124.

[23] Plichta, M. M., Heinzel, S., Ehlis, A. C., Pauli, P., Fallgatter, A. J. Model-based analysis of rapid event-related functional near-infrared spectroscopy (NIRS) data: a parametric validation study. Neuroimage. 2007;**35**(2):625–634.

[24] Schecklmann, M., Ehlis, A. C., Plichta, M. M., Fallgatter, A. J. Functional near-infrared spectroscopy: a long-term reliable tool for measuring brain activity during verbal fluency. Neuroimage. 2008;**43**(1):147–155.

[25] Okada, E., Delpy, D. T. Near-infrared light propagation in an adult head model. II. Effect of superficial tissue thickness on the sensitivity of the near-infrared spectroscopy signal. Applied Optics. 2003;**42**(16):2915–2921.

[26] Piper, S. K., Krueger, A., Koch, S. P., Mehnert, J., Habermehl, C., Steinbrink, J., Obrig, H., Schmitz, C. H. A wearable multi-channel fNIRS system for brain imaging in freely moving subjects. Neuroimage. 2014;**85**:64–71.

[27] Hiroyasu, T., Nakamura, Y., Yokouchi, H. Method for removing motion artifacts from fNIRS data using ICA and an acceleration sensor. In: 35th Annual International Conference of the IEEE Engineering in Medicine and Biology Society (EMBC); Jul 3; IEEE. 2013. pp. 6800–6803.

[28] Yamashita, Y., Maki, A., Koizumi, H. Wavelength dependence of the precision of non-invasive optical measurement of oxy-, deoxy-, and total-hemoglobin concentration. Medical Physics. 2001;**28**(6):1108–1114.

[29] Funane, T., Atsumori, H., Sato, H., Kiguchi, M., Maki, A. Relationship between wavelength combination and signal-to-noise ratio in measuring hemoglobin concentrations using visible or near-infrared light. Optical Review. 2009;**16**(4):442–448.

[30] Correia, T., Gibson, A., Hebden, J. Identification of the optimal wavelengths for optical topography: a photon measurement density function analysis. Journal of Biomedical Optics. 2010;**15**(5):056002.

[31] Corlu, A., Choe, R., Durduran, T., Lee, K., Schweiger, M., Arridge, S. R., Hillman, E. M., Yodh, A. G. Diffuse optical tomography with spectral constraints and wavelength optimization. Applied Optics. 2005;**44**(11):2082–2093.

[32] Zhu, T., Faulkner, S., Madaan, T., Bainbridge, A., Price, D., Thomas, D., Cady, E., Robertson, N., Golay, X., Tachtsidis, I. Optimal wavelength combinations for resolving in-vivo changes of haemoglobin and cytochrome-c-oxidase concentrations with NIRS. Biomedical Optics (Optical Society of America). 2012:JM3A–6.

[33] Scholkmann, F., Kleiser, S., Metz, A. J., Zimmermann, R., Mata Pavia, J., Wolf, U., Wolf, M. A review on continuous wave functional near-infrared spectroscopy and imaging instrumentation and methodology. Neuroimage. 2014;**85**(Part 1):6–27.

[34] Wolf, M., Ferrari, M., Quaresima, V. Progress of near-infrared spectroscopy and topography for brain and muscle clinical applications. Journal of Biomedical Optics. 2007;**12**(6):062104–062114.

[35] Custo, A., Wells III, W. M., Barnett, A. H., Hillman, E., Boas, D. A. Effective scattering coefficient of the cerebral spinal fluid in adult head models for diffuse optical imaging. Applied Optics. 2006;**45**(19):4747–4755.

[36] Boas, D., Culver, J., Stott, J., Dunn, A. Three dimensional Monte Carlo code for photon migration through complex heterogeneous media including the adult human head. Optics Express. 2002;**10**(3):159–170.

[37] Gibson, A., Hebden, J., Arridge, S. R. Recent advances in diffuse optical imaging. Physics in Medicine and Biology. 2005;**50**(4):R1.

[38] Virtanen, J., Noponen, T., Kotilahti, K., Virtanen, J., Ilmoniemi, R. J. Accelerometer-based method for correcting signal baseline changes caused by motion artifacts in medical near-infrared spectroscopy. Journal of Biomedical Optics. 2011;**16**(8):087005–087009.

[39] Bang, J. W., Choi, J. S., Park, K. R. Noise reduction in brainwaves by using both EEG signals and frontal viewing camera images. Sensors. 2013;**13**(5):6272–6294.

[40] Iramina, K., Matsuda, K., Ide, J., Noguchi, Y. Monitoring system of neuronal activity and moving activity without restraint using wireless EEG, NIRS and accelerometer. In: EEE EMBS Conference, editor. Biomedical Engineering and Sciences (IECBES); November; IEEE; 2010. pp. 481–484.

[41] Yücel, M. A., Selb, J., Boas, D. A., Cash, S. S., Cooper, R. J. Reducing motion artifacts for long-term clinical NIRS monitoring using collodion-fixed prism-based optical fibers. Neuroimage. 2014;**85**:192–201.

[42] Yücel, M. A., Selb, J., Cooper, R. J., Boas, D. A. Targeted principle component analysis: a new motion artifact correction approach for near-infrared Spectroscopy. Journal of Innovative Optical Health Sciences. 2014;**7**(02):1350066.

[43] Jin, Z. M., Yan, Y. X., Luo, X. J., Tao, J. W. A study on the dynamic pressure comfort of tight seamless sportswear. Journal of Fiber Bioengineering and Informatics. 2008;**1**(3):217–224.

[44] Kassab, A., Le Lan, J., Vannasing, P., Sawan, M. Functional near-infrared spectroscopy caps for brain activity monitoring: a review. Applied Optics. 2015;**54**(3):576–586.

[45] Ball, R., Shu, C., Xi, P., Rioux, M., Luximon, Y., Molenbroek, J. A comparison between Chinese and Caucasian head shapes. Applied Ergonomics. 2010;**41**(6):832–839.

[46] Atsumori, H., Kiguchi, M., Obata, A., Sato, H., Katura, T., Utsugi, K., Funane, T., Maki, A. Development of a multi-channel, portable optical topography system. In: 29th Annual International Conference of the IEEE Engineering in Medicine and Biology Society; August; IEEE. 2007. pp. 3362–3364.

[47] Kiguchi, M., Atsumori, H., Fukasaku, I., Kumagai, Y., Funane, T., Maki, A., Kasai, Y., Ninomiya, A. Wearable near-infrared spectroscopy imager for haired region. Review of Scientific Instruments. 2012;**83**(5):056101.

[48] Kassab, A. The design and development of a NIRS cap for brain activity monitoring [thesis]. Montreal, QC, Canada: Universite of Montreal; 2014.

[49] Brown, E., Rodenberg, N., Amend, J., Mozeika, A., Steltz, E., Zakin, M. R., Lipson, H., Jaeger, H. M. Universal robotic gripper based on the jamming of granular material. Proceedings of the National Academy of Sciences. 2010;**107**(44):18809–18814.

[50] Bicchi, A. Hands for dexterous manipulation and robust grasping: a difficult road toward simplicity. IEEE Transactions on Robotics and Automation. 2000;**16**(6):652–662.

Chapter 2

Near-Infrared Spectroscopy (NIRS): A Novel Tool for Intravascular Coronary Imaging

Marie-Jeanne Bertrand, Philippe Lavoie-L'Allier and
Jean-Claude Tardif

Additional information is available at the end of the chapter

http://dx.doi.org/10.5772/67196

Abstract

Acute coronary syndrome (ACS) arising from plaque rupture is the leading cause of mortality worldwide. Near-infrared spectroscopy (NIRS) combined with intravascular ultrasound (NIRS-IVUS) is a novel catheter-based intravascular imaging modality that provides a chemogram of the coronary artery wall, which enables the detection of lipid core and specific quantification of lipid accumulation measured as the lipid-core burden index (LCBI) in patients undergoing coronary angiography. Recent studies have shown that NIRS-IVUS can identify vulnerable plaques and vulnerable patients associated with increased risk of adverse cardiovascular events, whereas an increased coronary plaque LCBI may predict a higher risk of future cardiovascular events and periprocedural events. NIRS is a promising tool for the detection of vulnerable plaques in CAD patients, PCI-guidance procedures, and assessment of lipid-lowering therapies. Previous trials have evaluated the impact of statin therapy on coronary NIRS defined lipid cores, whereas NIRS could further be used as a surrogate end point of future ACS in phase II clinical trials evaluating novel anti-atheromatous drug therapies. Multiple ongoing studies address the different potential clinical applications of NIRS-IVUS imaging as a valuable tool for coronary plaque characterization and predictor of future coronary events in CAD patients.

Keywords: near-infrared spectroscopy (NIRS), intravascular ultrasound (IVUS), thin-cap fibroatheroma (TCFA), acute coronary syndrome (ACS), vulnerable plaque

1. Introduction

Coronary artery disease (CAD) is the leading cause of global mortality and the rupture of an unstable atherosclerotic plaque precedes the majority of acute coronary syndromes (ACS) [1, 2]. Autopsy studies have shown that the putative substrate for most ACS and many cases of sudden cardiac death (SCD) is the rupture of a thin-cap fibroatheroma (TCFA), the so-called *"vulnerable plaque,"* which is defined by a large lipid-rich necrotic core (NC) infiltrated with abundant macrophages and separated from the bloodstream by a thin fibrous cap [3, 4]. The ability to accurately detect index lesions using intravascular imaging is a potential attractive strategy, although it still remains a challenge in daily practice. Conventional coronary angiography (CCA) has been and continues to be an invaluable tool for epicardial coronary stenoses assessment and treatment [5]. Since the coronary angiogram provides a limited *"luminogram"* view of the coronary arteries, it cannot assess the properties of the arterial wall and thus tends to underestimate the true magnitude of plaque burden, especially in early stages of the disease in which positive vascular remodeling leads to a normal lumen caliber appearance on angiography despite substantial vascular wall plaque [6, 7]. Moreover, angiography provides no information in regard to plaque composition and biological activity, whereas intravascular imaging can potentially circumvent those limitations [8]. Several intravascular-imaging modalities, such as angioscopy, intravascular ultrasound (IVUS), virtual histology (VH), optical coherence tomography (OCT), and near-infrared spectroscopy (NIRS), have been developed throughout the quest of vulnerable plaque to characterize plaque composition and progression, to optimize patient risk stratification and for guiding therapy [9].

Near-infrared spectroscopy (NIRS) is a novel intravascular-imaging modality that provides chemical assessment related to the presence of cholesterol esters in lipid cores and

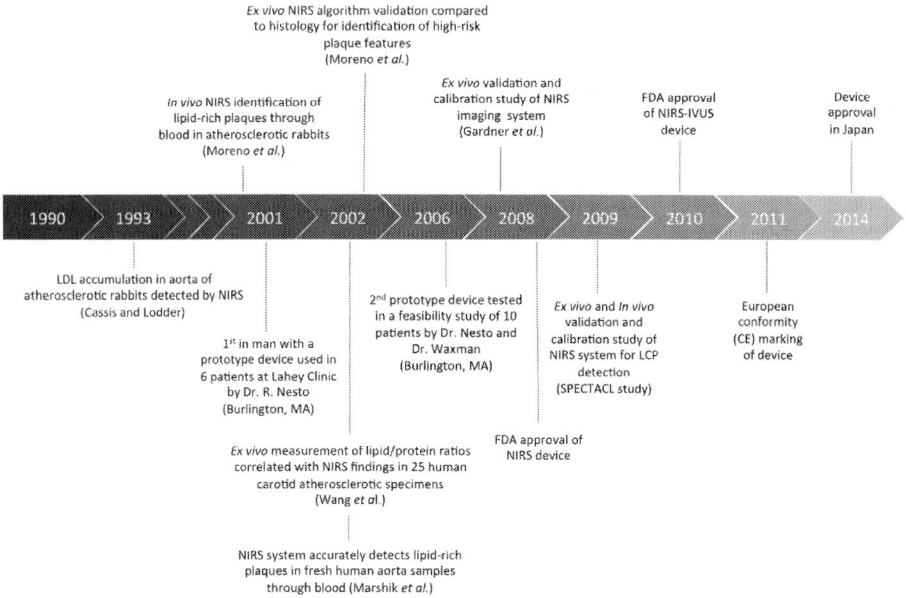

Figure 1. Timeline regarding important steps toward NIRS-IVUS imaging system development and use in clinical applications.

can generate spectra that distinguishes cholesterol from collagen in coronary plaques through their unique spectroscopic fingerprints [10]. NIRS was first used in 1993 for the detection of lipid content in an experimental animal model [11], followed by subsequent *ex vivo* validation in human cadavers [12]. In 2001, a device prototype for intracoronary imaging was developed, which led to multiple case series and clinical studies in the following decade [13–15]. This technology aims to detect vulnerable lipid-rich plaques (LRPs) by NIRS chemogram [16], whereas recent literature has demonstrated the association of LRP and culprit lesions in ACS [17, 18], as well as with nonculprit lesions in ACS [19], in percutaneous coronary intervention (PCI)-related procedural complications [20, 21], in plaque regression with statins therapy [22] and with the occurrence of cardiovascular events [23]. NIRS received US Food and Drug Administration (FDA) approval for clinical use in 2008 and for NIRS-IVUS system in 2010, followed by regulatory approval in Europe (CE marked) and Japan in 2011 and 2014, respectively (e.g., **Figure 1**) [24].

2. Near-infrared spectroscopy system

2.1. Principles of diffuse reflectance NIRS

Spectroscopy is based on the analysis of electromagnetic spectra induced by near-infrared light and provides direct evaluation of plaque composition, which could yield information on plaque vulnerability [13]. Several spectroscopic methods have been investigated for the purpose of identifying atherosclerotic plaque composition, although the commercially available catheter uses diffuse reflectance NIRS [13, 25]. The principle of NIRS relies on the interaction of light in the form of photons with different functional groups of organic molecules in a tissue, which results in reflected light in the NIR region from molecular vibrational energy in the form of oscillations of atoms within their chemical bonds. Photons can be absorbed or scattered by tissue, which determines the amount of light that is detected by the spectrometer. The wavelengths of light in NIRS are approximately in the 800–2500 nm range. Unique combinations of carbon-hydrogen (C-H), nitrogen-hydrogen (N-H), and oxygen-hydrogen (O-H) bonds are responsible for the major absorption of NIR light, whereas each functional group of large complex molecules yields absorption patterns at specific wavelengths, known as the *spectroscopic chemical fingerprint*, that provides qualitative and quantitative information on sample recognition and tissue characterization (e.g., **Figure 2**) [13, 26, 27].

Diffuse reflectance NIR spectroscopy has many features that enable *in vivo* lipid-core plaques (LCP) analysis in coronary arteries. The term "near" indicates the section of infrared that is closer to the visible light region with a longer wavelength and hence a lower energy than visible light. NIR has the ability to identify organic compounds from light penetration through blood and tissue, since hemoglobin and water have relatively low absorbance in the NIR wavelength, avoiding the need to be in contact with tissue or to clear the field of view with saline or contrast flush or by vessel occlusion [13, 26]. Moreover, it can provide simultaneous image acquisition and nondestructive chemical analysis of biologic tissue with rapid acquisition time (<1 s) from an ultrafast laser source, overcoming cardiac motion artifacts. Diffuse

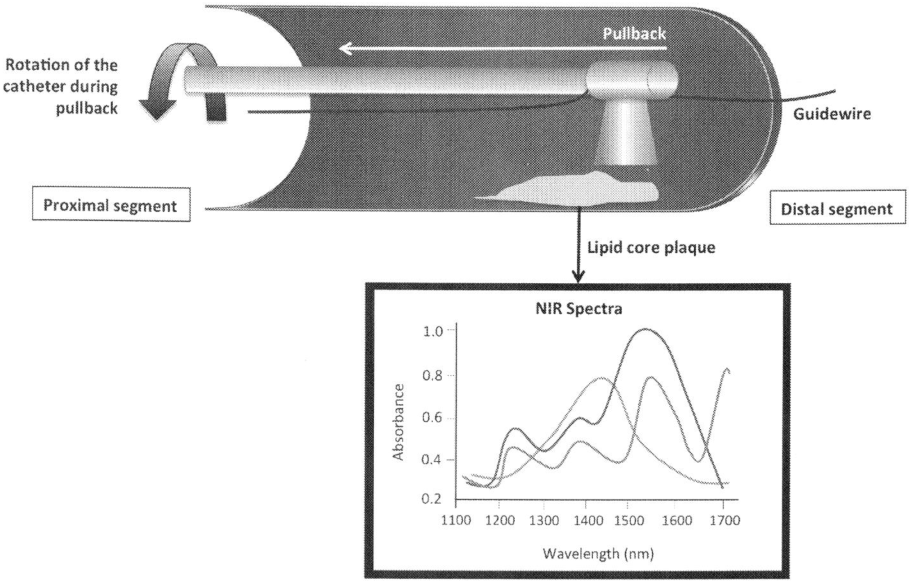

Figure 2. Near-infrared spectra detection and analysis of various components of a lipid-core plaque by NIRS-imaging system. NIRS intracoronary imaging is performed by the catheter's optical tip under automated rotating pullback that enables to rapidly scan the arterial vessel wall circumferentially and longitudinally. The catheter tip emits and collects light that interacts with different functional groups of molecules of the arterial wall and plots the relative absorbance of light across the wavelength range, which generates a spectrum. Thousands of NIR spectra are collected and produces a unique chemical "fingerprint" of the lipid-core plaque.

NIR spectroscopy has been used to identify multiple plasma constituents, to monitor systemic and cerebral oxygenation and also provides a specific chemical measure of LCP [13, 26, 27]. Other spectroscopy techniques are currently under research development for intravascular applications, including Raman spectroscopy, fluorescence spectroscopy, and magnetic resonance spectroscopy (e.g., **Table 1**) [13, 25].

2.2. NIRS-IVUS-combined catheter system

Spectroscopy has a strong fundamental basis for compositional measurement and is a highly efficient method for the identification of chemical components of unknown organic molecules. A single NIRS modality catheter system, the Lipiscan™ (InfraRedx Inc., Burlington, MA, USA), was first developed for invasive detection of LCP [26]. In order to obtain anatomical information on the vessel and optimal plaque characterization, a hybrid technology (TVC Imaging System™, InfraRedx Inc.) combining near-infrared spectroscopy(NIRS) and intravascular imaging (IVUS) was further developed, which allows simultaneous, co-registered acquisition of structural and compositional data of coronary artery plaques. Thus, combining the two complementary technologies enables a complete assessment of patient's arteries, including vessel size and structure, plaque volume, area, and composition [26, 35].

	Raman NIRS	Fluorescence spectroscopy	Diffuse reflectance NIRS	Nuclear magnetic resonance (NMR) spectroscopy
Principle	Raman shift from the scattering of a photon upon interaction with matter, generating a near-infrared wavelength forming the Raman spectra	Absorbance of energy from a tissue exposed to ultraviolet light, which in turns releases energy in the form of light	Reflected light from a tissue detected by the spectrometer at a wavelength, generating a NIR spectrum	Chemical shift from chemical groups exposed to an oscillating electromagnetic field and frequencies decoded by the Fourier transform to generate NMR spectrum
Plaque characterization	Cholesterol esters, collagen, phospholipids, triglycerides, calcium	Collagen, elastin fibers, lipoproteins, calcium, macrophages, foam cells	Lipid-core plaques	Unsaturated and polyunsaturated fatty acids, cholesterol esters, phospholipids, triglycerides
Validation studies	*Ex vivo* and *in vivo* animal and human studies	*In vitro* and *ex vivo* animal and human studies	*Ex vivo* and *in vivo* animal and human studies	13-Carbon NMR used in *ex vivo* and *in vivo* animal studies
Advantages	Evaluates the chemical composition of living tissues Signal more specific but weaker than diffuse reflectance NIRS (difficult to detect signal *in vivo*)	Strong fluorescence in arterial tissue, enabling rapid time acquisitions	Evaluates the chemical composition of living tissues, NIR light can penetrate blood and acquire signals from structures several millimeters deep relative to tissue surface	Lack of ionizing radiation (less radioactivity with carbon-13), noninvasive modality, enables to study several biological processes with metabolic, physiologic, and anatomic data combined to imaging
Availability	In development—fiber optics catheter-based system for PCI applications under investigation	No *in vivo* applications due to fluorescence signal distortion by hemoglobin	Catheter-based NIRS-IVUS system used as a clinical application	Costly, preclinical research

IVUS: intravascular ultrasound, NIRS: near-infrared spectroscopy; NMR: nuclear magnetic resonance; PCI: percutaneous coronary intervention [13, 28–34].

Table 1. Summary of different spectroscopic methods.

The commercially available NIRS-IVUS imaging system consists of a 3.2-French (F) rapid-exchange catheter compatible with 6F-guiding catheters, a pullback and rotation device, and a console that houses the scanning NIR laser, the computer that processes the spectral signal and two monitors [10, 26, 36]. Within the catheter body is a rotating core of optical fibers that deliver near-infrared light and measure the proportion of light reflected back over the range of optical wavelength (800–2500 nm) in the form of an imaging spectrum. The catheter-imaging core enables to collect data rapidly by rotating at 960 rpm with synchronized pullback at an automated speed of 0.5 mm/s. The system acquires >30,000 spectra per 100 mm. IVUS images are simultaneously acquired by a transducer at a frequency of 40 MHz and with an axial resolution of 100 μm, together with co-registered NIRS measurements, with a maximum imaging length of 12 cm and a depth of 1 mm or less. Thus, the NIRS spectra data are mapped and paired with corresponding cross-sectional IVUS frames, presented as a ring around the IVUS image [26, 27, 35, 36]. An upgrade version of the TVC catheter Imaging System™ was released by the company in 2015, which uses an extended bandwidth transducer that generates IVUS images at frequencies between 30 and 70 MHz, thus increasing the resolution and depth-to-field of the images [36].

2.3. Interpretation of NIRS data

Upon completion of the automated pullback scan, spectral data are automatically analyzed by a computer-based algorithm that transforms NIR spectra into a probability of LCP presence. The probability is mapped to a color pixel that will generate a digital two-dimensional color map of the artery called the NIRS chemogram, which represents the probability of the presence of LCP over the scanned segment of a vessel (**Figure 3**). On the longitudinal chemogram, the x-axis denotes the pullback location (in millimeters) and the y-axis represents the circumferential position (degrees of catheter rotation, from 0 to 360°). For each pixel of 0.1-mm length and 1° angle, the lipid-core probability is calculated from the spectral data collected and quantitatively coded on a color scale transitioning from red (0 = low probability of LCP) to yellow (high probability of LCP), with a probability of 0.60. The threshold required for the detection of LCP of interest was defined in the SPECTACL study according to the high prevalence of LCP (58%) detected in scanned segments that met both criteria of spectral adequacy and similarity from 60 patients undergoing PCI for stable CAD or ACS [10]. Pixels with intermediate data, including those that interfere with the guidewire, appear black. The block chemogram is a semi-quantitative summary metric of the probability that an LCP is present in a 2-mm NIRS chemogram segment that is computed and is displayed as a false color map, thus providing a 1:1 direct comparison of the chemogram with histopathology during validation of the lipid prediction algorithm. The blocks correspond to one of four colors (red ($P < 0.57$), orange ($0.57 \leq P < 0.84$), tan ($0.84 \leq P < 0.98$), and yellow ($P \geq 0.98$)), which represents the 90th percentile probability of lipid within the 2-mm segment of the pullback [26, 27, 35, 36]. The 2-mm block chemogram measures were used to compare the NIR spectra to histology in each 2-mm block in a receiver operating characteristic (ROC) curve analysis of diagnostic accuracy, from which LCP probabilities were calculated [10].

Chemometrics is the methodology applied by NIRS technology to analyze lipid content in atherosclerotic arteries [37]. The NIRS system was used in an extensive *ex vivo* study using human coronary arteries autopsy specimens to develop an algorithm for LCP detection. NIR spectra and histological data, used as gold standard, were collected from human autopsy hearts to build a calibration model capable of recognizing the NIR spectral shapes unique to LCP (see Section 2.4.2) [38]. Mathematical models constructed from a calibration set of samples were used to extract and analyze data from NIRS spectra, as reference values for the chemical compounds of interest in the tissue samples were obtained from histopathology samples. Models constructed from these calibration samples correlate the NIRS signals with

http://dx.doi.org/10.5772/67196

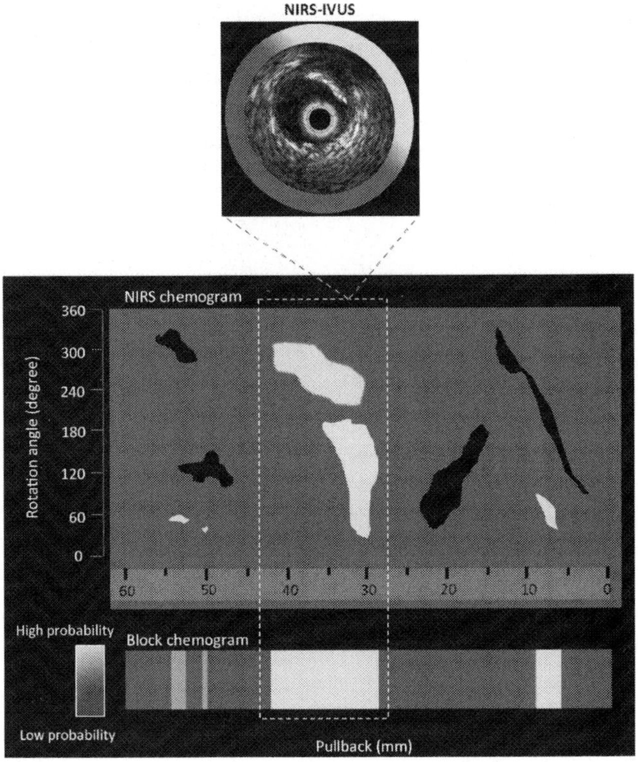

Figure 3. Example of a near-infrared spectroscopy (NIRS) chemogram. The near-infrared spectroscopy chemogram is a digital color-coded map of the arterial wall that is generated from NIR spectra analysis of the arterial wall, which indicates the location and intensity of lipid core in the region of interest (ROI). The *X*-axis represents the pullback position (in mm) and the *Y*-axis indicates the circumferential position of the measurement (in degrees). The block chemogram is a vertical summary of the chemogram at 2-mm pullback intervals. IVUS images are simultaneously acquired and co-registered with NIRS measurements and displayed as cross-section images superimposed with a chemogram.

the reference values, allowing the prediction of future samples on the basis of their NIRS measurements [39, 40]. The algorithm for LCP detection in humans was then prospectively validated in the SPECTACL study, in which chemograms obtained *in vivo* were similar to those obtained in histology controls (see Section 2.4.3) [10].

The lipid-core burden index (LCBI) is a measure of the lipid burden within the scanned region, calculated by dividing the number of yellow pixels that exceed an LCP probability of 0.6 per million by the total number of valid pixels in the segment, then multiplied by a factor of 1000 (LCBI range: 0–1000). Other measures can be computed on the chemogram image, such as the LCBI of a region of interest (ROI) and the maximum LCBI of the 4-mm region within the highest lipid burden within the ROI ($maxLCBI_{4mm}$) [26, 27, 35, 36, 39]. It has been shown that a high

LCBI detected in coronary plaques is associated with an increased risk of future cardiovascular events and periprocedural complications (see Section 2.6), which suggests that LCBI could be a useful biomarker for risk assessment and therapeutic efficacy in future clinical trials.

2.4. Validation of the NIRS-imaging system

2.4.1. Preclinical and autopsy studies

Autopsy, animal, and human studies have been carried out to test the utility and safety of NIRS for the purpose of eventually bringing this technology to patients in the catheterization laboratory. Cassis and Lodder first demonstrated the ability of NIRS to accurately identify low-density lipoprotein (LDL) *ex vivo* in the aorta of hypercholesterolemic rabbits [11, 41]. Furthermore, Jarros et al. [42] demonstrated that the cholesterol content of human aortic samples determined by NIR spectroscopy correlated strongly with that measured by reversed-phase, high-pressure liquid chromatography (correlation coefficient of 0.96). The ability of NIR spectroscopy to detect atherosclerosis in tissue was also demonstrated in human carotid and coronary arteries. Dempsey et al. [27] used diffuse reflectance NIR spectroscopy for the analysis of human carotid plaques exposed at the time of surgery. Transcutaneous NIRS was performed in the operating room during surgical endarterectomy and a NIRS algorithm was developed, using gel electrophoresis as a reference method, to determine lipoprotein composition in carotid specimen from NIR spectra. Their results showed significant near-IR correlation between certain lipoproteins present in carotid plaques and microscopic findings, including microscopic necrosis and ulceration, plaque hemorrhage, and thrombosis. Moreover, these proteins were easily detectable in patients with a medical history of CAD, coronary artery bypass grafting (CABG), and major surgery, and were also correlated with age, sex, and CAD risk factors. Furthermore, Wang et al. [12] reported that *ex vivo* direct measurement of lipid/protein ratios in human carotid atherosclerotic specimens from 25 patients correlated with NIRS spectroscopic findings. Thus, the authors concluded that these ratios could further be used to characterize advanced lesion types with superficial necrotic cores *in vivo* with NIR spectroscopy fitted with a fiber optic probe.

The first study to test the hypothesis that NIR spectroscopy could identify plaque composition and features associated with plaque vulnerability, defined by histology as the presence of lipid pool, thin fibrous cap (<65 μm by ocular micrometry), and inflammatory cell infiltration, was performed in 199 human aortic samples obtained at the time of autopsy [43]. An algorithm was constructed using NIR spectra obtained from 50% of the samples (calibration set) and was then tested on unknown samples (validation set) to determine its ability to identify high-risk features as determined by histology. Spectra associated with each of the three histological features of interest were defined by the results obtained from the calibration set. The main findings of this study were that NIRS could identify histology features associated with plaque vulnerability in human plaques *in vitro*, with a sensitivity and specificity of 90% (35 of 39 lesions) and 93% (56 of 60 lesions) for lipid pool, 77% (13 of 17 lesions) and 93% (76 of 82 lesions) for thin cap, and 84% (37 of 44 lesions) and 91% (49 of 55 lesions) for inflammatory cells, respectively. Moreno et al. [44] measured the NIRS spectra of 167 sections of fixed coronary artery samples and validated an algorithm against histology for the determination of lipid areas > or <0.6 mm^2, with a sensitivity and a specificity for lipid-rich coronary plaque detection of 83% (5 of 6 lesions) and 94% (60 of 64 lesions), respectively.

Since the intention of inventers of the NIRS system was to commercialize a catheter-based instrument that could assess plaques in coronary arteries *in vivo* and rapidly perform thousands of measurements through blood, Moreno et al. [45] first demonstrated that NIRS could identify lipid-rich plaques *in vivo* through blood in aorta of rabbits with diet-induced atherosclerosis. The catheter NIR spectroscopy was able to identify lipid areas > or <0.75 mm^2 with 78% sensitivity and 75% specificity. Marshik et al. [46] subsequently demonstrated accurate detection by NIRS spectra of lipid-rich plaques from 26 fresh human aorta samples through various amounts of blood up to a depth of 3 mm, with a sensitivity of 88% and a specificity of 79%. Moreover, the performance of the system was evaluated against histology, with favorable results for the detection of thin-cap fibroatheroma (TCFA) and disrupted plaques through blood, thus supporting the development of a NIR catheter for *in vivo* coronary arteries TCFA assessment [47]. To evaluate the performance of the system during cardiac motion, a human coronary autopsy specimen was attached at the surface of a beating pig's heart and connected to the porcine circulation [47]. The prototype 3.2-F NIRS catheter was positioned inside the coronary segment and was able to correctly identify a spectrally distinct target attached to the surface of the graft, despite blood flow and cardiac motion [48, 49].

2.4.2. Autopsy calibration and validation studies

The catheter-based system was improved with the addition of an automated pullback and rotation device allowing the system to circumferentially scan the length of a vessel. Calibration and validation studies of NIRS for the detection of LCP were first performed in human autopsy specimens of coronary arteries [16, 35]. The largest *ex vivo* study, conducted by Gardner et al. [38], aimed to evaluate the ability of the NIRS system to detect LCP in human coronary arteries from 84 autopsied hearts. Coronary arteries, obtained from a broad range of patient characteristics and causes of death, were mounted in a tissue fixture and connected to a blood circulation system with physiologic pressure, temperature, and flow. The resulting set of NIRS spectra and corresponding histology data were used to construct and validate an LCP detection algorithm. A total of 86 coronary segments from 33 hearts were used to calibrate the system algorithm for LCP detection and produced prospectively defined end points. The following 51 hearts and 126 segments were used to validate the accuracy of NIRS in the detection of LCPs in a double-blind, prospective study. In order to develop and validate the algorithm for the identification of LCP in coronary arteries, LCP of interest was defined as a fibroatheroma (FA) containing a lipid core of >0.2-mm thick, with a circumferential span of >60° on cross-section and a mean fibrous cap thickness of <450 μm. Prospective validation of the system for the detection of LCP from 51 hearts yielded an area under the ROC curve (AUC) of 0.80 (95% confidence interval (CI): 0.76–0.85) for average lumen diameters of up to 3.0 mm. The detection of any-sized fibroatheroma in an artery segment using the LCBI as a measure of lipid burden resulted in an AUC of 0.86 (95% confidence interval (CI): 0.81–0.91). However, false-positive scan results were obtained when the NIRS system was detecting areas with lipid that did not meet criteria of LCP. Moreover, LCPs with extensive calcifications were not detected by NIRS since the near-infrared light cannot penetrate through calcium and other artifacts [22, 38].

2.4.3. *Clinical validation studies*

The first use of the NIRS system in coronary arteries of living humans was performed in six patients undergoing elective PCI for stable angina using an early prototype (2001; Lahey Clinic, Burlington, MA) [13, 16, 40]. No device-related adverse events occurred, showing the safety and feasibility of the system to distinguish spectra measured through blood. However, significant motion artifacts were present due to slow-signal acquisition time (2.5 s). In August 2005, an improved ultrafast NIR system prototype was developed with a faster scanning laser and was later used in a feasibility study of 10 patients in 2006 (Lahey Clinic, Burlington, MA). The trial confirmed the safety of the newer improved device and showed its ability to discriminate between signals obtained in the artery and those from blood alone, with no measurable artifacts of motion [16, 40].

A subsequent pivotal study, the SPECTACL (SPECTroscopic Assessment of Coronary Lipid) clinical study, was performed to validate the accuracy of LCP-detected NIRS signals collected in coronary arteries of 106 patients [10]. The study met its primary end point of demonstrating that spectral data could be safely acquired in coronary arteries of patients with the intravascular NIRS system and that the spectra were equivalent to those gathered from autopsy specimens (success rate of 0.83; 95% confidence interval (CI): 0.70–0.93). Thus, this study supported the feasibility of LCP detection in living patients. Subsequent studies showed intra- and inter-catheter reproducibility of automated interpretation of NIR spectra signals [50, 51].

2.5. Comparison with other intravascular imaging modalities for plaque characterization

The most common cause of acute coronary syndromes (ACS) is believed to be coronary artery thrombosis due to the rupture of lipid-rich "*vulnerable plaques.*" Thin-cap fibroatheroma (TCFA) plaques, which are characterized by a lipid-laden necrotic core with an overlying thin fibrous cap measuring <65 μm, containing few smooth muscle cells but numerous macrophages, are often the substrate for plaque rupture-induced ACS [3, 4]. TCFAs are associated with positive remodeling and thus predominantly located in areas of the coronary tree that show mild to moderate luminal narrowing [52]. As previously outlined, coronary angiography only detects gross stenotic plaques and provides no insight regarding non-ruptured "*vulnerable plaques,*" which limits plaque burden assessment [6]. Intravascular imaging modalities have been developed to fill part of the gap in information provided by coronary angiography and for *in vivo* detection of LCP [35, 53]. *In vivo* atherosclerotic imaging could enable to detect, predict, and prevent plaque rupture, improve PCI treatment of flow limiting target lesions, and could identify new therapeutic targets that would prevent future adverse coronary events in CAD patients (e.g., **Table 2**).

http://dx.doi.org/10.5772/67196

	Spatial resolution (μm)	Depth (mm)	Energy source	Remodeling	Plaque composition	Calcium	Fibrous cap	Lipid core	Thrombus	Macrophages	Neovessels
IVUS	100–150	10	Ultrasound	++	–	++	±	+	±	–	–
RF-IVUS	100	10	Ultrasound	–	+	++	+	+	–	–	–
OCT	10	2–3	Near-infrared light	–	+	++	++	+	++	+	+
NIRS	1000	–	Near-infrared light	–	–	–	–	++	–	–	–
NIRS-IVUS	100–150	10	Near-infrared light + ultrasound	++	–	++	±	++	±	–	–

IVUS: intravascular ultrasound; NIRS: near-infrared spectroscopy; OCT: optical coherence tomography; RF-IVUS: radiofrequency intravascular ultrasound.

Table 2. A comparison of different intravascular imaging modalities.

2.5.1. *Intravascular ultrasound (IVUS) imaging*

Intravascular ultrasound imaging (IVUS) produces cross-sectional images of the lumen and the artery wall *in vivo,* enabling visual assessment of plaque echogenicity from axial resolution of approximately 100 μm using high-frequency detectors (up to 45 MHz) [9]. IVUS is very accurate in identifying calcifications (sensitivity and specificity of approximately 90%), plaque burden and, unlike coronary angiography, can detect non-protruding plaques as well as positive and negative vascular remodeling [9, 54]. Thus, IVUS is currently the gold standard for atherosclerotic imaging of the coronary arteries in progression/regression plaque trials [9, 55–57]. In addition to its use as a research tool, IVUS has shown to be of clinical value for the assessment of ambiguous lesions and facilitates optimal PCI procedures by providing reference vessel diameter [9, 58]. A previous study from Lee et al. [59] showed that attenuated lesions on IVUS were more common in ACS patients and were associated with more severe and complex plaque morphology, plaque burden, and higher frequency of no-reflow phenomenon during PCI procedures. Conventional grayscale IVUS has a high sensitivity for detecting lipid deposits (78–95%), visualized as echolucent zones, but a low specificity (30%) [54]. Another limitation of IVUS imaging is the low-axial resolution that does not allow to precisely define thin-cap fibroatheroma (TFCA), whose thickness is usually less than 65 μm in unstable plaques, and thus cannot identify plaques prone to rupture [54].

2.5.2. *Virtual histology (VH) imaging*

As compared to conventional invasive ultrasound techniques, radiofrequency (RF) IVUS provides additional information on plaque composition and morphology by spectral analysis of ultrasound backscatter [60]. A color-coded map allows the distinction of different components of atherosclerotic plaques, such as calcification (white), lipid/fibrofatty (light-green), fibrous (green) tissue, and necrotic core (red) [61]. Virtual histology (VH)-IVUS spectral analysis correlates with histopathology studies of plaques and can identify the four plaque components with sensitivity, specificity, and predictive accuracy ranging from 80 to 92% [54, 62, 63]. VH-IVUS detection of LCPs has been associated with higher incidence of clinical events [64, 65] and periprocedural complications during PCI [66–68]. Prospective assessment of vulnerable plaques was performed in the PROSPECT (Providing Regional Observations to Study Predictors of Events in the Coronary Tree) trial, a multicenter multimodality study that prospectively analyzed by IVUS and IVUS-VH imaging the coronary arteries of 697 ACS patients [64]. Their findings suggested that the presence of TCFA defined by VH-IVUS (hazard ratio (HR), 3.35; 95% CI, 1.77–6.36; $P < 0.001$), a minimal lumen area of ≤4 mm^2 (HR, 3.21; 95% CI, 1.61–6.42; $P = 0.001$), and a large plaque burden of ≥70% (HR, 5.03; 95% CI, 2.51–10.11; $P < 0.001$) were independent predictors of major adverse cardiovascular events (MACEs) in nonculprit lesions at 3.4 years follow-up. However, the positive-predictive value was only 18–23%, reflecting MACE's low prevalence. Although this study validated the concept of vulnerable plaque, the lack of specificity and difficulties in image interpretation/measurements prevented these results from changing clinical practice. The VIVA study [65], as well as the PREDICTION [69] and ATHEROREMO-IVUS [70] studies, subsequently reported similar findings, despite differences with the PROSPECT study regarding inclusion criteria, follow-up duration, definitions of TCFA and MACE. Although RF-IVUS is a validated and promising tool to identify patients and lesions at risk of future ACS, there are limitations regarding axial resolution, accuracy of necrotic core determination, and proper data acquisition and analysis [54, 64, 65].

2.5.3. *Optical coherence tomography (OCT)*

Optical coherence tomography is an invasive catheter-based imaging modality that measures the intensity and echo time delay of reflected near-infrared light from internal structures in tissues [71]. This technique provides a resolution of 10–20 μm *in vivo*, which is largely superior to IVUS. The recent technology uses the optical frequency domain imaging (OFDI), which enables faster pullback speeds without altering image quality and resolution [9]. The use of non-occlusive techniques with flushing of contrast through the guiding catheter during simultaneous image acquisition has partly resolved the issue of light absorption by blood components. OCT can discriminate features of high-risk plaques by evaluating the lipid content and macrophages infiltration, as well as the measurement of fibrous cap thickness [72]. This imaging modality is also used during percutaneous coronary intervention to assess stent apposition, coronary dissections, neoatherosclerosis and in-stent restenosis, mechanisms of plaque disruption in ACS patients, and more recently to evaluate the scaffold of bioabsorbable stents [73, 74]. The main limitation of OCT is the shallow penetration depth (1.0–2.5 mm) into the tissue, which limits proper imaging of biomarkers in atherosclerotic plaques [9, 75]. Other limitations include the lack of standardization of fibrous cap thickness analysis and the inconsistent accuracy in discriminating lipid-rich plaques from similar optical properties, such as macrophages accumulation, which can lead to false-positive results [72]. Regardless of the limitations, intracoronary FD-OCT remains a promising new clinical method for interrogating the microstructural details of the coronary wall [76].

2.5.4. *Near-infrared spectroscopy (NIRS)*

In contrast to IVUS, RF-IVUS, and OCT, which collect structural information, NIRS is unique for its ability to directly identify the chemical composition of the arterial wall and assess the presence of the LCP. NIRS detects unequivocal fingerprints from lipid core that is not affected by signal loss behind calcium due to acoustic shadowing, as it can occasionally preclude grayscale IVUS analysis, and the validation of NIRS included both calcified and non-calcified lipid cores in the definition of LCP [38]. NIRS alone does not provide information about structural anatomic parameters, such as vessel remodeling, plaque thickness, lumen area, and calcification [77]. However, as previously mentioned, the combined NIRS-IVUS-imaging catheter allows co-registration of both IVUS and NIRS data, which gives information on both plaque composition and morphology. NIRS-IVUS has shown to improve LCP detection, by comparison to IVUS, in calcified plaques as well as in lesions with small plaque burden [78]. The combined measures of plaque burden and LCBI improved the accuracy of fibroatheroma detection as compared with plaque burden alone by grayscale IVUS. Indeed, Puri et al. [79] conducted an *ex vivo* NIRS and IVUS-imaging study, performed in 116 coronary arteries of 51 autopsied hearts, whereas lesion-based analysis demonstrated that combining plaque burden and LCBI analysis significantly improves fibroatheroma detection accuracy (c index 0.77, $P = 0.028$), by comparison to plaque burden alone.

Several studies have compared NIRS with other intravascular-imaging modalities for LCP detection. It was previously shown that large plaque area measured by grayscale IVUS was more often associated with lipid accumulation/LCP detected by NIRS [19, 80]. However, Brugaletta et al. [80] found a weak correlation between the VH necrotic core content of the plaque and the block chemogram probability values (r =0.149), which did not improve after correction for the presence of calcium. In a larger study performed in 131 plaques of 66 vessels, in which 31 plaques (26.7%) were attenuated, the relation between VH-derived percentage necrotic core and NIRS-derived LCBI was not significant (r = 0.16, P = 0.110) [81]. However, after separation of the plaques according to grayscale IVUS morphology, a positive relationship between VH-derived maximum percentage necrotic core and LCBI was found in non-calcified plaques, but not in calcified plaques. A study conducted in 17 patients who underwent NIRS and OCT imaging showed modest linear correlation between LCBI and maximum lipid arc and lipid index measured by OCT (r^2 = 0.319, P = 0.003, and r^2 = 0.404, P = 0.001, respectively) [82]. Furthermore, Roleder et al. [83] conducted a study which aimed to evaluate the accuracy of NIRS-IVUS-imaging modality to detect TCFA in 60 patients with stable CAD, by comparison to OCT used as the gold-standard reference to define TCFA (cap thickness of <65 μm). They showed that OCT-defined TCFA was characterized by positive vessel remodeling with higher lipid-core burden, while NIRS revealed greater LCBI per 2-mm segment ($LCBI_{2mm}$) >315 with a remodeling index >1.046 as a combined criterion value.

In summary, there are important differences in LCP detection between different intravascular-imaging modalities, owing to their different imaging properties and limitations. As previously mentioned, OCT has the highest resolution but the weakest tissue penetration, limiting assessment of plaque burden and overall plaque volume [84]. While IVUS-VH and OCT require image interpretation for the detection of LCP, NIRS provides automated LCP detection without the need for manual imaging processing, facilitating its use in the catheterization laboratory and enabling rapid ad hoc clinical decision making during procedures. Moreover, OCT and NIRS can image through calcified lesions, whereas IVUS cannot. VH-IVUS can incorrectly misclassify intracoronary stents as calcium surrounded by necrotic core, a major limitation that is not found with OCT and NIRS imaging [84]. From the strengths and weaknesses of each individual imaging modality, it appears that the combination of two or more imaging technologies could improve LCP and vulnerable plaque detection [85].

2.6. NIRS-IVUS clinical applications

There is growing evidence from multiple studies of the clinical applications and value of the NIRS-IVUS imaging modality, including identifying the culprit lesion in ACS, optimizing PCI procedure, identifying plaques at high risk of periprocedural complications, for risk stratification, monitoring lipid-lowering therapy, and assessing plaque vulnerability (e.g., **Table 3**) [86].

2.6.1. In vivo detection of culprit lesions in ACS

Several studies have evaluated NIRS detection of LCP, shown by an increased LCBI, at the site of culprit lesions associated with coronary events. Madder et al. [17] performed NIRS imaging in culprit vessels of 20 patients with acute ST-segment elevation myocardial infarc-

http://dx.doi.org/10.5772/67196

Setting	Study or authors	Publishing year	*N*	Clinical end point(s)	Results, references
LCP detection and *in vivo* validation of NIRS imaging	SPECTACL	2009	106	(1) Evaluate the similarities of NIRS spectra obtained in patients to spectra previously obtained and validated by histology in autopsy specimens; (2) to assess the safety of the device; and (3) to quantify the presence of LCP at target and non-target sites	NIRS system enables to safely obtain spectral data in patients that were similar to those from autopsy specimens and results demonstrated the feasibility of invasive detection of coronary LCP [10]
Plaque characterization	Brugaletta et al.	2011	31	Compare the findings of NIRS, IVUS-VH and IVUS grayscale obtained in matched coronary vessel segments of patients undergoing coronary angiography	Larger plaque area by grayscale IVUS was more often associated with either elevated percentage of VH derived-necrotic core (NC) or LCP by NIRS; correlation between LCP detected by NIRS and NC by VH was weak [80]
	Pu et al.	2012	66	Evaluate NIRS combined with IVUS to provide novel information on human coronary plaque characterization	Combining NIRS and IVUS contributes to plaque characterization *in vivo* [81]
Vulnerable plaque	ATHEROREMO-NIRS	2014	203	Determine the long-term prognostic value of intracoronary NIRS as assessed in a nonculprit vessel in patients with CAD	CAD patients with an LCBI ≥ 43.0 had a fourfold risk of MACE during 1-year follow-up [92]
	Madder et al.	2016	121	Evaluate the association between large lipid-rich plaques (LRP) detected by NIRS at non-stented sites in a target artery and subsequent MACCE	Detection of large LRP by NIRS ($maxLCBI_{4mm}$ ≥400) at non-stented sites in a target vessel was associated with an increased risk of future MACCE [93]
Acute coronary syndrome	Madder et al.	2012	60	Determine the frequency of LCP at target and remote sites in ACS vs. stable angina patients	Target lesions responsible for ACS were frequently composed of LCP; LCP in culprit and non culprit lesions were more common in patients with ACS vs. stable angina patients [77]
	Madder et al.	2013	20	To describe NIRS findings in culprits lesions of STEMI patients	maxLCBI4mm > 400 detected in vivo by NIRS is a threshold for identification of STEMI culprit plaques [17]
	Madder et al.	2015	81	Assess the lipid burden of culprit lesions in NSTEMI and UA patients	LCP similar to those detected at STEMI culprit sites were detected at culprit sites of NSTEMI and UA patients [18]

Setting	Study or authors	Publishing year	*N*	Clinical end point(s)	Results, references
Periprocedural MI	COLOR registry	2011	62	Analyse the relationship between the presence of large LCP detected by NIRS and periprocedural MI	NIRS provides a rapid and automated detection of extensive LCPs that are associated with a high risk of periprocedural MI [20]
	Raghunathan et al.	2011	30	Evaluate if an association between the presence and extend of LCP detected by NIRS before PCI and periprocedural MI	PCI of LCP detected lesions by NIRS is associated with increased risk of MI after PCI [21]
	Maini et al.	2013	77	Evaluation of LCP modification with coronary revascularization and its correlation with periprocedural MI	Plaque modification can be performed by interventional methods and evaluated with NIRS; axial plaque shifting is an acute prognostic marker for postprocedural MI [124]
PCI optimization	Dixon et al.	2012	69	Compare the target lesions length using NIRS combined with angiography vs. angiography alone	Patients undergoing stent implantation could have LCP extended beyond angiographic margins of the initial target lesion using QCA alone [97]
	Hanson et al.	2015	58	Assess the prevalence of plaque burden and LCP extended beyond angiographic borders of target lesions	NIRS-IVUS imaging demonstrates that target lesion length is commonly underestimated by angiography alone [98]
	Ali et al.	2013	65	Characterize neointimal composition of in-stent restenosis in both BMS and DES using a multimodality approach with OCT and NIRS-IVUS	In-stent thin-cap neoatherosclerosis is more prevalent, more diffusely distributed across stented segment and is associated with increased periprocedural MI in DES compared with BMS [108]
	Madder et al.	2016	120	Evaluate NIRS-IVUS system findings of increased lipid signals in pre-existing stents, speculated to indicate neoatherosclerosis, and compare with a control group of freshly implanted stents, in which any lipid signal originates from fibroatheroma under the stent	Detection of LCP in pre-existing stents by NIRS alone is not reliable evidence of neoatherosclerosis, as the lipid signal may originate from fibroatheroma under the stent [109]
Monitoring lipid-lowering therapies	YELLOW	2013	87	Determine the impact of short-term intensive statin therapy (Rosuvastatin 40 mg OD) on intracoronary plaque content	Short-term intensive treatment with statin was associated with a significant reduction in LCBI / lipid content compared to standard therapy [22]

http://dx.doi.org/10.5772/67196

Setting	Study or authors	Publishing year	*N*	Clinical end point(s)	Results, references
Prevention of PCI complications	Brilakis et al.	2012	9	Investigate whether the use of a distal embolic protection device might prevent complications of LCP interventions	The use of a distal protection device frequently resulted in embolized material retrieval after stenting of native coronary artery lesions with large LCP [123]
	CANARY	2015	85	Evaluate if a distal protection device reduce postprocedural MI for PCI of LCP lesions	Distal protection device dis not reduce postprocedural MI [125]
	Erlinge et al.	2015	18	Evaluate if aspiration thrombectomy reduces the lipid content of culprit plaques by NIRS-IVUS in ACS patients assessed	Thrombus aspiration resulted in a 28% reduction in lipid content by performing aspiration thrombectomy in culprit lesion [129]

ACS: acute coronary syndrome; BMS: bare-metal stent; CAD: coronary artery disease; DES: drug-eluting stent; IVUS: intravascular ultrasound; LCBI: lipid-core burden index; LCP: lipid-core plaque; LRP: lipid-rich plaque; MACE: major adverse cardiac events; MACCE: major adverse cardiac and cerebrovascular events; MI: myocardial infarction; NC: necrotic core; NIRS; near-infrared spectroscopy; NSTEMI: non-ST-segment elevation myocardial infarction OCT: optical coherence tomography; PCI: percutaneous coronary intervention; QCA: quantitative coronary angiography; STEMI: ST-segment elevation myocardial infarction; UA: unstable angina; VH: virtual histology.

Table 3. A summary of intracoronary human NIRS clinical studies to identify lipid-core plaques (LCPs).

tion (STEMI) and compared their findings with spectra analysis in nonculprit segments of the artery and with autopsy control segments. The $maxLCBI_{4mm}$ was 5.8-fold higher in STEMI culprit segments than in 87 nonculprit segments of the STEMI culprit vessel (median (interquartile range (IQR)): 523 [445 to 821] vs. 90 [6 to 265]; $P < 0.001$). Moreover, $maxLCBI_{4mm}$ was 87-fold higher than in 279 coronary autopsy segments free of large LCP by histology (median (interquartile range (IQR)): 523 [445 to 821] vs. 6 [0 to 88]; $P < 0.001$). Thus, a threshold of $maxLCBI_{4mm}$ ≥400 distinguished STEMI culprit segments *in vivo* from coronary artery autopsy segments free of LCP with high accuracy (sensitivity: 85%; specificity: 98%) [17]. Among the first 85 STEMI cases, two patients showed culprit lesions that did not contain lipid plaque, but rather a calcified nodule in one case and a coronary dissection in the other [15].

Similar NIRS findings of lipid burden were observed in culprit lesions of patients in non-ST segment elevation myocardial infarction (NSTEMI) [18, 77]. LCPs are more common in patients with ACS compared to stable angina patients. From the 81 NSTEMI and unstable angina (UA) patients who underwent culprit vessel NIRS imaging prior to stenting, non-STEMI culprit segments had a 3.4-fold greater $maxLCBI_{4mm}$ than nonculprit segments (448 ± 229 vs. 132 ± 154, $P < 0.001$) and unstable segments had a 2.6-fold higher $maxLCBI_{4mm}$ than nonculprit lesions (381 ± 239 vs. 146 ± 175, $P < 0.001$) [18]. Culprit segments in NSTEMI patients were more often characterized by a $maxLCBI_{4mm}$ ≥400 than those with UA, with a sensitivity of 63.6% versus 38.5%, respectively. Moreover, a large LCP was identified by NIRS within the culprit lesions of five cases of resuscitated out-of-hospital cardiac arrest that subsequently underwent coronary angiography [87]. There is a stepwise increase in lipid content, represented by $maxLCBI_{4mm}$, from nonculprit lesions (0–130), to unstable angina (≈380), to NSTEMI (≈450) and STEMI patients (≈550), supporting the concept of more fibrotic lesions in stable angina and more lipid-rich vulnerable plaque in STEMI, NSTEMI, and sudden death [15]. NIRS-IVUS evidence of LCP with a large plaque burden suggests that the lesion is a culprit, and that such information could be relevant in patients with ambiguous coronary angiography for efficient treatment management.

2.6.2. *Association with cardiovascular risk factors*

A recent clinical study has evaluated the association between clinical risk factors and blood characteristics of vascular inflammation and lipid content/LCP visualized by NIRS. de Boer et al. [19] reported the use of NIRS in a nonculprit coronary artery in 208 patients undergoing percutaneous coronary intervention or invasive diagnostic coronary exploration for various indications. It was found that male gender, hypercholesterolemia, and the presence of multivessel CAD were modestly associated with higher LCBI values on NIRS. A history of peripheral vascular disease and/or cerebral disease and the use of beta-blockers were positively associated with LCBI, while biomarkers such as blood lipids and high-sensitivity C-reactive protein were not. All clinical characteristics reflecting patients with high CAD risk explained only 23% of the variability in LCBI. Moreover, the LCBI on NIRS and the percentage area of plaque burden determined by IVUS were modestly correlated (r =0.29). In the light of these results, this study could not address the prognosis value of NIRS-imaging modality. Methodological caveats could in part explain the low correlation obtained between NIRS and

http://dx.doi.org/10.5772/67196

IVUS imaging, including the use of lower-frequency IVUS catheters (20 MHz), IVUS and NIRS acquisitions performed using different catheters, measurement of a single cross section on IVUS, and the absence of data regarding the reproducibility of repeated NIRS pullbacks and measurements [88].

2.6.3. *Assessing plaque vulnerability and risk stratification*

Retrospective autopsy studies have revealed specific histological culprit lesion morphologies in patients suffering from an ACS, which has created an enthusiasm in the use of intravascular coronary artery imaging in search of the *"vulnerable plaque"* at risk of rupture and endoluminal thrombosis. The thin-cap fibroatheroma (TCFA) is believed to be the precursor lesion of plaque rupture, although there is a lack of prospective robust evidence in the literature [15, 89]. A prospective animal study conducted in an atherosclerotic and diabetic pig model showed that NIRS-IVUS imaging can detect and predict the future development of inflamed fibroatheromas with subsequent validation against postmortem histology [90]. The features of rupture-prone plaques included thinned fibrous cap, increased plaque and necrotic core areas, increased concentration of activated inflammatory cells, and the presence of apoptotic and proliferating cells within the fibrous cap [90]. An autopsy study of 103 coronary arteries from 56 autopsied hearts, aiming to assess grayscale IVUS and NIRS detection of histological fibroatheroma (FA), with histology validation as the gold standard, showed that both superficial IVUS attenuation and NIRS-LCP had a similar high specificity of approximately 95% in detecting FAs, however IVUS showed a low sensitivity (36% vs. 47%; P =001) [91]. The addition of NIRS significantly increased the accuracy of fibroatheroma detection at the minimum lumen area from 75% to 89% among all cross-sections ($P < 0.05$). When either IVUS attenuation or lipid-rich plaque was present, the sensitivity for prediction of an FA was significantly higher compared with IVUS alone (63% vs. 36%, $P < 0.001$) and NIRS alone (84% vs. 65%, $P < 0.001$).

The first prospective human study, published in 2014, has evaluated the association of high LCP by NIRS and cardiovascular events. The ATHEROREMO-NIRS (The European Collaborative Project on Inflammation and Vascular Wall Remodeling in Atherosclerosis—Near-Infrared Spectroscopy) trial is a prospective, observational study that aimed to evaluate the prognostic value of NIRS in a nonculprit coronary artery from 203 patients referred for angiography due to stable angina or ACS [92]. The results showed that the 1-year cumulative incidence of all-cause mortality, non-fatal ACS, stroke, and unplanned coronary revascularization was 4-fold increased in patients with an LCBI equal or above to the median value of 43.0 compared to those with an LCBI value below the median (adjusted HR: 4.04; 95% CI: 1.33–12.29; $P = 0.01$). The association of the LCBI value with primary end point was similar in both stable and ACS patients. Although these results are promising, the number of events in this trial was small, and therefore studies with larger number of events will be required for the validation of vulnerable patient detection with NIRS-IVUS imaging. A more recent NIRS-IVUS single-center registry study was conducted in 121 consecutive patients undergoing combined NIRS and IVUS imaging to evaluate the association of large lipid-rich plaques at non-stented sites in a target vessel and subsequent major adverse cardiovascular and cerebrovascular events (MACCE) [93]. The results showed that the presence of large LCP in a non-stented segment, defined by NIRS $maxLCBI_{4mm}$ ≥400 at baseline, was associated with a significantly increased risk of future MACCE during follow-up (HR 10.2, 95% CI: 3.4–30.6; $P < 0.001$). This study, although single center, underpowered, and with limited follow-up, was consistent with the findings of ATHEROREMO-NIRS study, whereas NIRS detection of lipid burden was associated with patient-level risk of future MACCE [93].

The detection of fibroatheroma could help to identify culprit lesions in ACS patients, predict lesions subject to periprocedural complications, could allow optimal stent selection, and reduce the rate of stent restenosis. Whether the detection of fibroatheroma using NIRS-IVUS will prevent future events is currently being studied in several trials, including the Lipid-Rich Plaque study (LCP; Clinical Trials.org Identifier: NCT02033694), PROSPECT II ABSORB trial (A Multicentre Prospective Natural History Study Using Multimodality Imaging in Patients With acute Coronary Syndromes; Clinical Trials.org Identifier: NCT02171065), and ORACLE-NIRS trial (Lipid cORe Plaque Association With CLinical Events: a Near-InfraRed Spectroscopy Study; Clinical Trials.org Identifier: NCT02265146).

2.6.4. *Optimizing percutaneous coronary intervention procedures*

Visual estimation of a coronary stenosis on a two-dimensional (2D) angiography or quantitative coronary angiography (QCA) of lesion lengths is often misleading from image foreshortening and underestimation of plaque burden. IVUS offers accurate length measurement during automated pullback, proximal and distal reference diameter of a vessel, and enables to evaluate the presence and extent of calcifications [26]. The ADAPT-DES (Assessment of Dual Antiplatelet Therapy With Drug-Eluting Stents) study, a prospective, multicenter, nonrandomized "all-comers" registry of 8583 consecutive patients, showed that IVUS-guidance PCI, performed in 39% of patients, was associated with reduced 1-year rates of MACE (3.1% vs. 4.7%; adjusted HR, 0.70; 95% CI: 0.55–0.88; P = 002), as compared to angiography guidance alone [94]. The benefits of IVUS were observed in patients with ACS and complex lesions, although significant reductions in MACE were present in all patient subgroups, including stable angina and single-vessel disease. Similar results were observed in subsequent meta-analysis [95, 96].

The use of combined NIRS-IVUS imaging may further optimize stent implantation by accurate identification of lipid margins, and thus cover all the segments with high lipid burden. Dixon et al. [97] analyzed 75 lesions with NIRS imaging and demonstrated that lipid-core plaque extended beyond the angiographic margins of the initial target lesion in 16% of cases. Hanson et al. [98] showed that atheroma, defined as plaque burden >40% or LCP, extended beyond angiographic margins in 52 of the 58 lesions analyzed with NIRS-IVUS (90% of lesions), with a mean lesion length that was significantly longer when assessed by NIRS-IVUS as compared with angiography alone (19.8 ± 7.0 vs. 13.4 ± 5.9 mm; $P < 0.0001$). Those results suggests that NIRS-IVUS guidance during PCI procedures, as a "red-to-red" stenting strategy, could optimize complete LCP coverage by a stent with the proper length according to the landing zones and thus reduce the risk of edge dissections, stent failure, and subsequent adverse clinical outcomes [26, 39, 99–101]. Although it seems rationale to implant the edges of a stent in a normal artery segment, the marginal increased risk of stent thrombosis and restenosis with longer stents will require future studies to determine if routine use of NIRS-IVUS for proper stent sizing will improve patient outcomes [102].

Detection of lipid core in a lesion has also been used as one of the factors to consider in the decision to implant a bare metal stent (BMS) or a drug-eluting stent (DES). Several studies have demonstrated a greater frequency of stent thrombosis after DES implantation when struts were penetrating into a lipid-rich necrotic core plaque rather than in a non-yellow (fibrous) plaque [103, 104]. The absence of struts coverage by the formation of a neointima layer during vessel's healing process was seen with both DES and BMS implantation in lipid-rich plaques, which is likely the underlying mechanism of stent thrombosis seen in those patients [105, 106]. Neoatherosclerosis is an important contributor to late-stent thrombosis with newer generation DES, as well as late in-stent restenosis. Histologically, neoatherosclerosis is characterized by the accumulation of lipid-laden macrophages within the neointima with or without necrotic core formation and/or calcification and can occur months to years following stent placement [107]. Originally described in postmortem studies, neoatherosclerosis has more recently been detected by intracoronary imaging. Ali et al. [108] used NIRS and OCT to assess the development of neoatherosclerosis in 65 consecutive patients with symptomatic in-stent restenosis. The prevalence of LCP within neointimal hyperplasia segments was 89% using NIRS versus 62% using OCT. Neoatherosclerosis was associated with significantly reduced minimal cap thickness with plaque rupture occurring exclusively in those patients. Moreover, DES had a higher prevalence and earlier occurrence of neoatherosclerosis, thinner cap, and more lipid burden and density. However, LCP identified by NIRS alone was not associated with periprocedural MI during treatment for in-stent restenosis, which reflects the limited ability of NIRS to differentiate lipid located within the neointimal tissue from a lipid core located underneath stent struts. Nevertheless, postmortem imaging and subsequent histology analysis showed that NIRS could correctly characterize lipid despite the presence of metal struts. Similar findings were reported in a study published by Madder et al. [109], whereas NIRS was not reliable for neoatherosclerosis detection when used as the sole imaging modality for LCP detection. The NIRS lipid signal could not distinguish neoatherosclerosis from fibroatheroma underlying the stent. No doubt that NIRS can detect coronary LCP, but it seems unlikely suitable as a standalone technique for accurate neoatherosclerosis detection and that the adjunction of IVUS or OCT will be required to determine the position of NIRS lipid signal relative to the underlying stent struts [110].

It was proposed that the growth of neointima tissue on the top of a vulnerable plaque might increase the thickness of the fibrous cap [103, 110, 111]. Brugaletta et al. [112] reported the ability of bioresorbable vascular scaffold (BVS) implantation to promote the growth of neointimal tissue, which acts as a barrier to isolate vulnerable plaques. An ongoing trial, the PROSPECT II ABSORB sub-study trial (Clinical Trials.org Identifier: NCT021711065), will randomize patients with plaques at high risk of causing future coronary events (plaque burden ≥70%) to receive an Absorb™ BVS (Abbott Vascular, IL, USA) with optimal medical therapy (OMT) versus OMT alone. This sub-study aims to evaluate the changes in the plaque at 2 years follow-up. Clinically, large LCPs have been shown to be associated with MACE, especially periprocedural myocardial infarction [21]. Whether lipid burden influences long-term outcomes following stent implantation remains elusive.

2.6.5. *Prevention of periprocedural complications*

Approximately 3–15% of percutaneous coronary interventions are complicated by periprocedural myocardial infarction (PPMI) and no-reflow, in part by distal embolization of intraluminal thrombus and/or lipid-core plaque content, which is associated with adverse long-term outcomes [113, 114]. It was reported that periprocedural MIs are associated with increased atherosclerotic burden and large LCPs [115–118]. Indeed, embolization of the lipid core after stent implantation in a plaque with high lipid content has been identified as an important cause of periprocedural no-reflow and MI with and without the presence of intracoronary thrombus [118–120]. A pilot study performed in nine patients using an embolic protection device showed that embolized material consisted in fibrin and platelet aggregates, which reflects the highly thrombogenic content of necrotic core of large atheroma plaques and LCP [98, 120, 121]. In a sub-study of the COLOR (Chemometric Observation of Lipid-Core Plaques of Interest in Native Coronary Arteries) registry, a prospective multicenter observational study aiming to determine a relationship between NIRS-defined high LCBI and periprocedural MI, Goldstein et al. [20] analyzed the cardiac biomarkers of 62 stable patients undergoing PCI. The main findings were that periprocedural MI, defined in the study as a postprocedural elevation above three times the upper limit of normal (ULN) for either creatine kinase-MB (CK-MB) or cTnI measured 4–24 h after PCI, occurred in nine patients (14.5%) and was more common among patients with a $maxLCBI_{4mm} \geq 500$ (7 of 14 patients, 50%) versus patients with a $maxLCBI_{4mm} < 500$ (2 of 48 patients, 4.2%). The authors concluded that a high LCP, defined as a $maxLCBI_{4mm} \geq 500$, was associated with periprocedural events. These results are concordant with the registry study conducted by Raghunathan et al. [21], in which the analysis of 30 patients who underwent pre-procedure NIRS imaging showed a postprocedural increase of CK-MB more than three times the UNL in 27% of patients with a ≥1 yellow blocks ($n = 11$) as opposed to none in the 19 patients without a yellow block within the stented lesion.

Distal embolization, as an important mechanism of periprocedural MI, was further supported by several studies that have demonstrated a significant decrease in the size of LCP after stenting [122–124]. Stone et al. showed in the CANARY trial that LCP measured as LCBI by NIRS in the stented vessels reduces with PCI treatment, with a significant reduction of median LCBI from 143.2 before PCI to 17.9 after PCI ($P < 0.001$) [125]. Moreover, the authors showed that the occurrence of periprocedural MI was associated with higher LCBI, results that are concordant with previous findings [20, 21].

In order to prevent periprocedural MI during PCI, several strategies were proposed during stenting procedures, including aspiration thrombectomy, embolization distal-protection devices, vasodilators, intensive anticoagulation, and antiplatelet therapies. The CANARY (Coronary Assessment by NIR of Atherosclerotic Rupture-Prone Yellow) trial randomized 85 stable angina patients undergoing stent implantation of a single native coronary lesion and pre-procedure NIRS-defined $maxLCBI_{4mm} \geq 600$ to PCI with or without distal-protection filter [125]. Among the 31 randomized cases with a $maxLCBI_{4mm} \geq 600$, there was no difference in the rates of periprocedural MI with or without the use of distal-protection filter (35.7 vs. 23.5%, respectively; relative risk 1.52; 95% CI: 0.50–4.60, P =0.69). It should be noted that the CANARY trial was ended prematurely due to difficulties in identifying patients suitable for randomization to embolic-protection devices and lack of signs of benefits and thus was not adequately powered to detect a difference in MI or other major procedural complications between the two patient groups. An ongoing study, the CONCERTO (Randomized-Controlled Trial of a Combined versus Conventional Percutaneous Intervention for Near-Infrared Spectroscopy Defined High-Risk Native Coronary Artery Lesions; Clinical Trials.org Identifier: NCT02601664) trial, aims to evaluate different strategies for periprocedural MI prevention. Patients undergoing PCI with high-risk native coronary lesion, defined as ≥2 contiguous yellow blocks on the block chemogram, are randomized to combined preventive measures versus conventional PCI. The combined preventive measures consist of pre-PCI administration of an intracoronary vasodilator and a glycoprotein IIb/IIIa inhibitor, in addition to the use of an embolic-protection device if technically feasible and a complete coverage of the LCP if technically feasible.

Thrombectomy is often used to aspirate thrombus and restore blood flow in the culprit vessel during primary PCI in STEMI patients. The clinical benefits of routine thrombus aspiration remain a matter of debate, since the TAPAS (Thrombus Aspiration during Percutaneous Coronary Intervention in Acute Myocardial Infarction) study demonstrated a reduction of mortality while larger studies such as TASTE (Thrombus Aspiration in ST-Elevation Myocardial Infarction in Scandinavia) and TOTAL (Trial of Routine Aspiration Thrombectomy with PCI versus PCI Alone in Patients with STEMI) did not show a reduction of cardiovascular mortality, with an increased rate of stroke at a 30-day follow-up in the TOTAL trial [126–128]. Erlinge et al. [129] performed NIRS-IVUS imaging in 18 ACS patients to examine if aspiration thrombectomy reduced the lipid content of ACS culprit plaques. The culprit lipid content was quantified by NIRS-IVUS before and after thrombectomy as the lipid-core burden index (LCBI), and aspirates were examined by histological staining for lipids, calcium, and macrophages. Culprit lesions were found to have high lipid content prior to thrombectomy, which resulted in a 28% reduction in culprit lesion lipid content (pre-aspiration LCBI 466 ± 141 vs. post-aspiration 335 ± 117, $P = 0.0001$).

As aforementioned, the use of intracoronary NIRS-IVUS imaging for accurate identification of LCP lesions prone to embolize, as well as different treatment strategies, for periprocedural MI prevention are attractive approaches, however their clinical benefits on myocardial salvage and prevention of embolization remains to be demonstrated in future studies.

2.6.6. Monitoring effects of lipid-lowering therapies

It is well known that statin therapy reduces rates of cardiovascular events in secondary prevention. The pharmacological effects of specific lipid-reducing agents that reduce free and esterified cholesterol could be evaluated with NIRS, as it informs on the lipid content of coronary artery plaques over time. The demonstration of markedly reduced LCBI values in a patient after 1 year of high-dose rosuvastatin therapy was the first indication that NIRS-IVUS could be used to evaluate the effect of systemic anti-atherosclerotic medical therapy [130]. In the YELLOW (Reduction in Yellow Plaque by Aggressive Lipid-Lowering Therapy) trial, Kini et al. [22] prospectively randomized 87 patients with multivessel coronary artery disease undergoing PCI with one culprit and one nonculprit hemodynamically significant lesions, defined by fractional flow reserve (FFR <0.80), to receive intensive statin therapy (rosuvastatin of 40 mg daily) or standard lipid-lowering therapy. The nonculprit lesions had a baseline assessment by NIRS-IVUS and FFR, prior to randomization. Rosuvastatin therapy resulted in a significant reduction in the plaque lipid content/maxLCBI$_{4mm}$ compared to standard therapy. The significant reduction in maxLCBI$_{4mm}$ associated with intensive statin therapy was observed across subgroups of the study population, based on age, gender, presence of diabetes, and baseline lipid profile. However, no significant changes were observed for the maxLCBI$_{4mm}$ and LCBI measurements at the lesion site in the standard lipid treatment group at follow-up. Although baseline LCBI was significantly higher in patients randomly allocated to intensive versus standard therapy, the YELLOW trial highlights that LCP measured by NIRS was associated with CAD and that it could be a potential tool to monitor regression of the disease in phase II clinical trials evaluating novel anti-atheromatous therapies.

A similar study of the effect of rosuvastatin treatment on the coronary plaque composition and necrotic core, the IBIS-3 (Integrated Biomarker and Imaging Study 3) trial, failed to demonstrate a significant reduction of necrotic core volume or LCBI under intensive rosuvastatin therapy for 1 year [131]. The effects of high-dose statin therapy are being further investigated in the YELLLOW II trial (Clinical Trials.org Identifier: NCT01837823), a phase II clinical study, that aims to assess the regression of plaque lipid content and changes in plaque morphology from atherosclerotic lesions after 8–12 weeks of high-dose statin therapy by utilizing NIRS, IVUS, and OCT imaging modalities in the coronary arteries.

2.7. Limitations of the technology

Near-infrared spectroscopy (NIRS) identifies the chemical signature of the lipid component, specifically lipid core-containing coronary plaque (LCP). The main limitations of NIRS technology are the lack of information regarding the lumen, plaque anatomy, and status of the fibrous cap or its attenuation. Although NIRS may be one of the most sensitive modalities to detect lipid-core plaques, it cannot provide information on the depth of the lipid core. Moreover, the accurate measurement of lipid volume/burden with NIRS has not been validated [132]. To overcome these pitfalls, a new combined imaging catheter adding intravascular ultrasound (IVUS) imaging was developed. However, since intravascular ultrasound has a low sensitivity to visualize lipid inside a plaque, the additional value of this new system will require further evaluation [26].

The clinical relevance of imaging specific features of the vulnerable plaque for risk stratification and clinical decision making remains unclear. Higher-resolution imaging modalities, such as OCT, better assessed determinants of vulnerable plaques than NIRS; however, there is currently no commercialized system combining OCT and NIRS modalities. The prognostic utility and incremental value of NIRS when associated with biomarkers of plaque vulnerability assessed by IVUS (plaque burden, MLA, and remodeling) remains to be investigated [26, 133]. Many studies have brought evidence that IVUS-guided PCI achieves superior outcomes compared to angiography guidance alone [134]. The potential value of adding NIRS for lipid-rich plaques at risk of embolization and for a complete coverage of LCPs remains to be investigated.

NIRS-IVUS-imaging modality is an invasive diagnostic modality that targets patients in the setting of secondary prevention, thus precluding its utilization for primary prevention, along with other invasive imaging technologies.

2.8. Future trials and perspective

NIRS-IVUS-imaging technology is improving and should become a sensitive modality for coronary plaque characterization. A new algorithm for collagen detection has been developed using the same spectroscopy signal, which enables to detect the amount of fibrous tissue over the LCP (thin or thick fibrous cap) [15]. This technology will be further optimized by adding a recently developed, but not yet available, high-resolution IVUS, which will allow to accurately differentiate between thin and thick fibrous caps. Co-registration of NIRS with other imaging modalities is also being developed. The use of combined OCT-NIRS catheters has been recently demonstrated as a proof of concept [15].

NIRS-IVUS has also been used in the carotid arteries to detect LCP, which could represent a suitable imaging modality to determine the risk of stroke or the risk of complications during carotid stent placement or endarterectomy. However, this new clinical application remains to be validated in future studies [15].

Multiple prospective outcome studies are currently ongoing to evaluate the ability of NIRS-IVUS imaging to detect vulnerable plaques that are likely to cause future adverse events. Among those studies are the LRP trial (Lipid-rich Plaque Study; Clinical Trial.org Identifier: NCT02033694), the PROSPECT II ABSORB trial (Providing Regional Observations to Study Predictors of Events in the Coronary Tree II; Clinical Trial.org Identifier: NCT02171065), and the ORACLE-NIRS trial (Lipid-core plaque association with clinical events: a near-infrared spectroscopy study; Clinical Trial.org Identifier: NCT02265146). The YELLOW II trial (NCT01837823), which aims to evaluate the effects of rosuvastatin treatment on lipid content after 8–10 weeks of treatment regimen, has completed patient enrolment but results are still pending. Another trial has been completed and awaiting for results publication, the NIRS-TICAGRELOR trial (Clinical Trial.org Identifier: NCT02282332), which aims to evaluate the effect of the P2Y12 inhibitor ticagrelor (AstraZeneca, Cambridge, England) on plaque stabilization and reduction of inflammation by NIRS-defined reduction of LCBI in patients on long-term statin therapy undergoing non-urgent PCI.

3. Conclusion

NIRS is a promising tool for the detection of vulnerable plaques in CAD patients, PCI-guidance procedures, and assessment of lipid-lowering therapies. NIRS-IVUS has been shown to be a reliable and reproducible modality for the detection of intracoronary LCPs, with validation using the current gold-standard, histology. It has already been shown that this imaging modality is highly specific for identifying NSTEMI and STEMI culprit plaques, that it can be used to follow the progression of vulnerable plaques over time, and to evaluate the effect of lipid-lowering therapies and intracoronary devices. Moreover, preliminary data have shown that NIRS-IVUS-imaging technology can identify vulnerable patients. Multiple ongoing clinical trials will hopefully validate this tool for vulnerable plaque and patient detection, as well as for treatment management and follow-up of patients with CAD.

Abbreviations

ACS	Acute coronary syndrome
BMS	Bare-metal stent
BVS	Bioresorbable vascular scaffold
CAD	Coronary artery disease
CABG	Coronary artery bypass graft
CCA	Conventional coronary angiography
CK-MB	Creatine kinase-MB
cTnI	Cardiac troponin I
DES	Drug-eluting stent
Fr	French
FA	Fibroatheroma
FDA	US Food and Drug Administration
FD-OCT	Frequency-domain optical coherence tomography
FFR	Fractional flow reserve
IVUS	Intravascular ultrasound
LCBI	Lipid-core burden index
LCP	Lipid-core plaque
LDL	Low-density lipoprotein
LRP	Lipid-rich plaque
MACE	Major adverse cardiac events
MACCE	Major adverse cardiac and cerebrovascular events
$maxLCBI_{4mm}$	Maximum lipid-core burden index in 4-mm region
MI	Myocardial infarction
MLA	Minimal lumen area
NC	Necrotic core
NIRS	Near-infrared spectroscopy

http://dx.doi.org/10.5772/67196

NSTEMI	Non-ST segment elevation myocardial infarction
OCT	Optical coherence tomography
OFDI	Optical frequency domain imaging
OMT	Optimal medical therapy
PCI	Percutaneous coronary intervention
PPMI	Periprocedural myocardial infarction
QCA	Quantitative coronary angiography
ROI	Region of interest
SCD	Sudden cardiac death
STEMI	ST-segment elevation myocardial infarction
TCFA	Thin-cap fibroatheroma
UA	Unstable angina
ULN	Upper limit of normal
VH	Virtual histology

Author details

Marie-Jeanne Bertrand[1,2], Philippe Lavoie-L'Allier[1,2] and Jean-Claude Tardif[1,2*]

*Address all correspondence to: jean-claude.tardif@icm-mhi.org

1 Montreal Heart Institute, Montreal, Canada

2 Faculty of Medicine, Université de Montréal, Montreal, Canada

References

[1] White HD, Chew DP. Acute myocardial infarction. Lancet. 2008;372:570–584. DOI: 10.1016/S0140-6736(08)61237-4

[2] Bentzon JF, Otsuka F, Virmani R, Falk E. Mechanisms of plaque formation and rupture. Circ Res. 2014;114:1852–66. DOI: 10.1161/CIRCRESAHA.114.302721

[3] Muller JE, Tofler GH, Stone PH. Circadian variation and triggers of onset of acute cardiovascular disease. Circulation. 1989;79:733–743. DOI: 10.1161/01.CIR.79.4.733

[4] Virmani R, Kolodgie FD, Burke AP, Farb A, Schwartz SM. Lessons from sudden coronary death: a comprehensive morphological classification scheme for atherosclerotic lesions. Arterioscler Thromb Vasc Biol. 2000;20:1262–1275. DOI: 10.1161/01.ATV.20.5.1262

[5] Chamuleau SA, van Eck-Smit BL, Meuwissen M, Piek JJ. Adequate patient selection for coronary revascularization: an overview of current methods used in daily clinical practice. Int J Cardiovasc Imaging 2002;18:5–15. DOI: 10.1023/A:1014372125457

[6] Mintz GS, Painter JA, Pichard AD, Kent KM, Satler LF, Popma JJ, Chuang YC, Bucher TA, Sokolowicz LE, Leon MB. Atherosclerosis in angiographically "normal" coronary artery reference segments: an intravascular ultrasound study with clinical correlations. J Am Coll Cardiol. 1995;25:1479–1485. DOI: 10.1016/0735-1097(95)00088-L

[7] Glagov S, Weisenberg E, Zarins C, Stankunavicius R, Kolletis G. Compensatory enlargement of human atherosclerotic coronary arteries. N Engl J Med. 1987;316:1371–1375. DOI: 10.1056/NEJM198705283162204.

[8] Goldstein JA. Angiographic plaque complexity: the tip of the unstable plaque iceberg. J Am Coll Cardiol. 2002;39:1464–1467. DOI: 10.1016/S0735-1097(02)01772-2

[9] Tardif JC, Lesage F, Harel F, Romeo P, Pressacco J. Imaging biomarkers in atherosclerosis trials. Circ Cardiovasc Imaging. 2011;4:319–333. DOI: 10.1161/CIRCIMAGING.110.962001

[10] Waxman S, Dixon SR, L'Allier P, Moses JW, Peterson JL, Cutlip D, Tardif JC, Nesto RW, Muller JE, Hendricks MJ, Sum ST, Gardner CM, Goldstein JA, Stone GW, Krucoff MW. *In vivo* validation of a catheter-based near-infrared spectroscopy system for detection of lipid core coronary plaques: initial results of the SPECTACL study. JACC Cardiovasc Imaging. 2009;2:858–868. DOI: 10.1016/l.jcmg.2009.05.001

[11] Cassis LA, Lodder RA. Near-IR imaging of atheromas in living arterial tissue. Anal Chem. 1993;65:1247–1256. DOI: 10.1021/ac00057a023

[12] Wang J, Geng YJ, Guo B, Klima T, Lal BN, et al. Near-infrared spectroscopic characterization of human advanced atherosclerotic plaques. J Am Coll Cardiol. 2002;39:1305–1313. DOI: 10.1016/S0735-1097(02)01767-9

[13] Moreno PR, Muller JE. Identification of high-risk atherosclerotic plaques: a survey of spectroscopic methods. Curr Opin Cardiol. 2002;17:638–647. ISSN: 0268-4705

[14] Jaguszewski M, Klingerberg R. Intracoronary near-infrared spectroscopy (NIRS) imaging for detection of lipid content of coronary plaques: current experience and future perspectives. Curr Cardiovasc Imaging Rep. 2013;6:426–430. DOI:10.1007/s12410-013-9224-2

[15] Erlinge D. Near-infrared spectroscopy for intracoronary detection of lipid-rich plaques to understand atherosclerotic plaque biology in man and guide clinical therapy. J Intern Med. 2015;278:110–125. DOI: 10.1111/joim.12381

[16] Caplan JD, Waxman S, Nesto RW, Muller JE. Near-infrared spectroscopy for the detection of vulnerable coronary artery plaques. J Am Coll Cardiol. 2006;47:C92–96. DOI: 10.1016/j.jacc.2005.12.045

[17] Madder RD, Goldstein JA, Madden SP, Puri R, Wolski K, Hendricks M, Sum ST, Kini A, Sharma S, Rizik D, Brilakis ES, Shunk KA, Petersen J, Weisz G, Virmani R, Nicholls SJ, Maehara A, Mintz GS, Stone GW, Muller JE. Detection by near-infrared spectroscopy of

large lipid core plaques at culprit sites in patients with acute ST-segment elevation myocardial infarction. JACC Cardiovasc Interv. 2013;6:838–846. DOI:10.1016/j.jcin.2013.04.012

[18] Madder RD, Husaini M, Davis AT, Van Oosterhout S, Harnek J, Götberg M, Erlinge D. Detection by near-infrared spectroscopy of large lipid cores at culprit sites in patients with non-ST-segment elevation myocardial infarction and unstable angina. Catheter Cardiovasc Interv. 2015;86:1014–1021. DOI: 10.1002/ccd.25754

[19] de Boer SPM, Brugaletta S, Garcia-Garcia HM, Simsek C, Heo JH, Lenzen MJ, Schultz C, Regar E, Zijlstra F, Boersma E, Serruys PW. Determinants of high cardiovascular risk in relation to plaque-composition of a non-culprit coronary segment visualized by near-infrared spectroscopy in patients undergoing percutaneous coronary intervention. Eur Heart J. 2014;35:282–289. DOI: 10.1093/eurheartj/eht378

[20] Goldstein JA, Maini B, Dixon SR, Brilakis ES, Grines CL, Rizik DG, Powers ER, Steinberg DH, Shunk KA, Weisz G, Moreno PR, Kini A, Sharma SK, Hendricks MJ, Sum ST, Madden SP, Muller JE, Stone GW, Kern MJ. Detection of lipid-core plaques by intracoronary near-infrared spectroscopy identifies high risk of periprocedural myocardial infarction. Circ Cardiovasc Interv. 2011;4:429–437. DOI: 10.1161/CIRCINTERVENTIONS.111.963264

[21] Raghunathan D, Abdel-Karim A-RR, Papayannis AC, daSilva M, Jeroudi OM, Rangan BV, Banerjee S, Brilakis ES. Relation between the presence and extent of coronary lipid core plaques detected by near-infrared spectroscopy with postpercutaneous coronary intervention myocardial infarction. Am J Cardiol. 2011;107:1613–1618. DOI: 10.1016/j.amjcard.2011.01.044

[22] Kini AS, Baber U, Kovacic JC, Limaye A, Ali ZA, Sweeny J, et al. Changes in plaque lipid content after short-term intensive versus standard statin therapy: the YELLOW trial (reduction in yellow plaque by aggressive lipid-lowering therapy). J Am Coll Cardiol. 2013;62:21–29. DOI: 10.1016/j.jacc.2013.03.058

[23] Oemrawsingh RM, Cheng JM, Garcia-Garcia HM, van Geuns R-J, de Boer SPM, Simsek C, et al. Near-infrared spectroscopy predicts cardiovascular outcome in patients with coronary artery disease. J Am Coll Cardiol. 2014;64:2510–2518. DOI: 10.1016/j.jacc.2014.07.998

[24] Jang IK. Near infrared spectroscopy: another toy or indispensable diagnostic tool? Circ Cardiovasc Interv. 2012;5:10–11.DOI: 10.1161/CIRCINTERVENTIONS.111.967935

[25] Jaffer FA, Verjans JW. Molecular imaging of atherosclerosis: clinical state-of-the-art. Heart 2014;100:1469–1477. DOI: 10.1136/heartjnl-2011-301370

[26] Kilic ID, Caiazzo G, Fabris E, Serdoz R, Abou-Sherif S, Madden S, Moreno PR, Goldstein J, Di Mario C. Near-infrared spectroscopy-intravascular ultrasound: scientific basis and clinical applications. Eur Heart J. 2015;16:1299–1306. DOI:10.1093/ehjci/jev208

[27] Dempsey RJ, Davis DG, Buice RG, Lodder RA. Biological and medical applications of near-infrared spectroscopy. Appl Spectrosc OSA. 1996;50:18A–34A. DOI: 10.1366/0003702963906537

[28] Downes A, Elfick A. Raman spectroscopy and related techniques in biomedicine. Sensors. 2010;10:1871–89. DOI: 10.3390/s100301871.

[29] Hanlon EB, Manoharan R, Koo TW, Shafer KE, Motz JT, Fitzmaurice M, Kramer JR, Itzkan I, Dasari RR, Feld MS. Prospects for in vivo Raman spectroscopy. Phys Med Biol. 2000;45:R1–59. DOI: 10.1088/0031-9155/45/2/201

[30] de Lima CJ, Sathaiah S, Silveira L, Zângaro RA, Pacheco MT. Development of catheters with low fiber background signals for Raman spectroscopic diagnosis applications. Artif Organs. 2000;24:231–234. DOI: 10.1046/j.1525-1594.2000.06525.x

[31] Marcu L, Fishbein MC, Maarek JM, Grundfest WS. Discrimination of human coronary artery atherosclerotic lipid-rich lesions by time-resolved laser-induced fluorescence spectroscopy. Arterioscler Thromb Vasc Biol. 2002;21:1244–1250. DOI: 10.1161/hq0701.092091

[32] Toussaint JF, Southern JF, Fuster V, Kantor HL.13C-NMR spectroscopy of human atherosclerotic lesions. Relation between fatty acid saturation, cholesteryl ester content, and luminal obstruction. Arterioscler Thromb. 1994;14:1951–1957. DOI: 10.1161/01.ATV.14.12.1951

[33] Peng S, Guo W, Morrisett JD, Johnstone MT, Hamilton JA. Quantification of cholesteryl esters in human and rabbit atherosclerotic plaques by magic-angle spinning ^{13}C-NMR. Arterioscler Thromb Vasc Biol. 2000;20:2682–2688. DOI: 10.1161/01.ATV.20.12.2682

[34] Trouart TP, Altbach Mi, Hunter GC, Eskelson CD, Gmitro AF. MRI and NMR spectroscopy of the lipids of atherosclerotic plaque in rabbits and humans. Magn Res Med. 1997;38:19–26. DOI: 10.1002/mrm.1910380105

[35] Shydo B, Hendricks M, Frazier G. Imaging of plaque composition and structure with the TVC Imaging System™ and TVC Insight™ catheter. J Invasive Cardiol. 2013;25:5A–8A. ISSN: 1557-2501

[36] Negi SI, Didier R, Ota H, Magalhaes MA, Popma CJ, Kollmer MR, Spad M-A, Torguson R, Suddath W, Satler LF, Pichard A, Waksman R. Role of near-infrared spectroscopy in intravascular coronary imaging. Cardiovasc Revasc Med. 2015;16:299–305. DOI: 10.1016/j.carrev.2015.06.001

[37] Lavine BK, Workman J. Chemometrics. Anal Chem. 2013;85:705–714. DOI: 10.1021/ac303193j

[38] Gardner CM, Tan H, Hull EL, Lisauskas JB, Sum ST, Meese TM, Jiang C, Madden SP, Caplan JD, Burke AP, Virmani R, Goldstein J, Muller JE. Detection of lipid core coronary plaques in autopsy specimens with a novel catheter-based near-infrared spectroscopy system. JACC Cardiovasc Imaging. 2008;1:638–648. DOI: 10.1016/j.jcmg.2008.06.001

[39] Danek BA, Karatasakis A, Madder RD, Muller JE, Madden S, Banerjee S, Brilakis ES. Experience with the multimodality near-infrared spectroscopy/intravascular ultrasound coronary imaging system: principles, clinical experience, and ongoing studies. Curr Cardiovasc Imaging Rep. 2016;9:7. DOI: 10.1007/s12410-015-9369-2

[40] Sum ST, Madden SP, Hendricks MJ, Chartier SJ, Muller JE. Near-infrared spectroscopy for the detection of lipid core coronary plaques. Curr Cardiovasc Imaging Rep. 2009;2:307–415. DOI: 10.1007/s12410-009-0036-3

[41] Lodder RA, Cassis L, Ciurczak EW. Arterial analysis with a novel near-IR fiber-optic probe. Spectroscopy. 1990;5:12–17.

[42] Jarros W, Neumeister V, Lattke P, Schuh D. Determination of cholesterol in atherosclerotic plaques using near infrared diffuse reflection spectroscopy. Atherosclerosis. 1999;147:327–337. DOI: 10.1016/S0021-9150(99)00203-8

[43] Moreno PR, Lodder RA, Purushothaman KR, Charash WE, O'Connor WN, Muller JE. Detection of lipid pool, thin fibrous cap, and inflammatory cells in human aortic atherosclerotic plaques by near-infrared spectroscopy. Circulation 2002;105:923–927. DOI:10.1161/hc0802.104291

[44] Moreno PR, Ryan SE, Hopkins D, Wise B, Purushothaman KR, Charash WE, O'Connor W, Muller JE. Identification of lipid-rich plaques in human coronary artery autopsy specimens by near-infrared spectroscopy. J Am Coll Cardiol. 2001;37:356A. DOI: 10.1016/S0735-1097(01)80005-X

[45] Moreno PR, Ryan SE, Hopkins D. Identification of lipid-rich aortic atherosclerotic plaques in living rabbit with a near infrared spectroscopy catheter. J Am Coll Cardiol. 2001;37:3A. DOI: 10.1016/S0735-1097(01)80001-2

[46] Marshik B, Tan H, Tang J, Lindquist A, Zuluaga A. Discrimination of lipid-rich plaques in human aorta specimens with NIR spectroscopy through whole blood. Am J Cardiol. 2002;90:129H. DOI: 10.1016/S0002-9149(02)02727-3

[47] Marshik B, Tan H, Tang J, Lindquist A, Zuluaga A. Detection of thin-capped fibroatheromas in human aorta tissue with near infrared spectroscopy through blood. J Am Coll Cardiol. 2003;41:42. DOI :10.1016/S0735-1097(03)80181-X

[48] Waxman S, Tang J, Marshik BJ, Tan H, Khabbaz KR, Connolly RJ, Dunn TA, Zuluaga AF, DeJesus S, Caplan JD, Muller EJ. In vivo detection of a coronary artificial target with a near infrared spectroscopy catheter.Am J Cardiol. 2004;94:141E. DOI: 10.1016/j.amjcard.2004.07.055

[49] Waxman S, Khabba K, Connolly R. Intravascular imaging of atherosclerotic human coronaries in a porcine model: a feasibility study. Int J Cardiovasc Imaging. 2008;24:37–44. DOI: 10.1007/s10554-007-9227-7

[50] Garcia BA, Wood F, Cipher D, Banerjee S, Brilakis ES. Reproducibility of near-infrared spectroscopy for the detection of lipid core coronary plaques and observed changes after coronary stent implantation. Catheter Cardiovasc Interv. 2010; 76:359–365. DOI: 10.1002/ccd.22500

[51] Abdel-Karim A-RR, Rangan B V, Banerjee S, Brilakis ES. Intercatheter reproducibility of near-infrared spectroscopy for the in vivo detection of coronary lipid core plaques. Catheter Cardiovasc Interv. 2011; 77:657–661. DOI: 10.1002/ccd.22763

[52] Kolodgie FD, Burke AP, Farb A, Gold HK, Yuan J, Narula J, Finn AV, Virmani R. The thin-cap fibroatheroma: a type of vulnerable plaque: the major precursor lesion to acute coronary syndromes. Curr Opin Cardiol. 2001;16:285–292. ISSN: 0268-4705

[53] Rizik D, Goldstein JA. NIRS-IVUS imaging to characterize the composition and structure of coronary plaques. J Invasive Cardiol. 2013;25:2A–4A. ISSN: 1557-2501

[54] Garcia-Garcia HM, Serruys PW. Advances in the invasive diagnosis and treatment of vulnerable coronary plaques. Eur Cardiol Rev 2008;4:105–110. ISSN: 0268-4705

[55] Nissen SE. Halting the progression of atherosclerosis with intensive lipid lowering: results from the Reversal of Atherosclerosis with Aggressive lipid Lowering (REVERSAL) trial. Am J Med. 2005;118:22–27. DOI: 10.1016/j.amjmed.2005.09.020

[56] Nissen SE, Nicholls SJ, Sipahi I, Libby P, Raichlen JS, Ballantyne CM, Davignon J, Erbel R, Fruchart JC, Tardif JC, Schoenhagen P, Crowe T, Cain V, Wolski K, Goormastic M, Tuzcu EM. Effect of very high-intensity statin therapy on regression of coronary atherosclerosis: The asteroid trial. JAMA. 2006;295:1556–65. DOI: 10.1001/jama.295.13.jpc60002

[57] Nicholls SJ, Ballantyne CM, Barter PJ, Chapman J, Erbel RM, Libby P, Raichlen JS, Uno K, Borgman M, Wolski K, Nissen SE. Effect of two intensive statin regimens on progression of coronary disease. N Engl J Med. 2011;365:2078–2087. DOI: 10.1056/NEJMoa1110874

[58] Lee SY, Mintz GS, Kim SY, Hong YJ, Kim SW, Okabe T, Pichard AD, Satler LF, Kent KM, Suddath WO, Waksman R, Weissman NJ. Attenuated plaque detected by intravascular ultrasound clinical, angiographic, and morphologic features and post-percutaneous coronary intervention complications in patients with acute coronary syndromes. JACC Cardiovasc Interv. 2009;2:65–72. DOI: 10.1016/j.jcin.2008.08.022

[59] Guedes A, Tardif JC. Intravascular ultrasound assessment of atherosclerosis. Curr Atheroscler Rep. 2004;6:219–224. DOI: 10.1007/s11883-004-0035-4

[60] Vince DG, Dixon KJ, Cothren RM, Cornhill JF. Comparison of texture analysis methods for the characterization of coronary plaques in intravascular ultrasound imaging. Comput Med Imaging Graph 2000;24:221–229. DOI: 10.1016/S0895-6111(00)00011-2

[61] Kawasaki M, Takatsu M, Noda T, Ito Y, Kunishima A, Arai M, Nishigaki K, Takemura G, Morita N, Minatoguchi S, Fujiwara H. Noninvasive quantitative tissue characterization and two-dimensional color-coded map of human atherosclerotic lesions using ultrasound integrated backscatter: comparison between histology and integrated backscatter images. J Am Coll Cardiol. 2001;38:486–492. DOI: 10.1016/S0735-1097(01)01393-6

[62] Nasu K, Tsuchikane E, Katoh O, Vince G, Virmani R, Surmely JF, Murata A, Takeda Y, Ito T, Ehara M, Matsubara T, Terashima M, Suzuki T. Accuracy of in vivo coronary plaque morphology assessment: a validation study of in vivo virtual histology compared with in vitro histopathology. J Am Coll Cardiol. 2006;47:2405–2412. DOI: 10.1016/j.jacc.2006.02.044

[63] Van Herck G, De Meyer G, Ennekens G, Van Herck P, Herman A, Vrints C. Validation of in vivo plaque characterisation by virtual histology in a rabbit model of atherosclerosis. EuroIntervention 2009;5:149–156.

[64] Stone GW, Maehara A, Lansky AJ, de Bruyne B, Cristea E, Mintz GS, Mehran R, McPherson J, Farhat N, Marso SP, Parise H, Templin B, White R, Zhang Z, Serruys PW. A prospective natural-history study of coronary atherosclerosis. N Engl J Med 2011;364:226–235. DOI: 10.1056/NEJMoa1002358

[65] Calvert PA, Obaid DR, O'Sullivan M, Shapiro LM, McNab D, Densem CG, Schofield PM, Braganza D, Clarke SC, Ray KK, West NE, Bennett MR. Association between IVUS finding and adverse outcomes in patients with coronary artery disease. The VIVA (VH-IVUS in vulnerable atherosclerosis) study. J Am Coll Cardiol Imaging 2011;8:894–901. DOI: 10.1016/j.jcmg.2011.05.005

[66] Kawaguchi R, Oshima S, Jingu M, Tsurugaya H, Toyama T, Hoshizaki H, Taniguchi K. Usefulness of virtual histology intravascular ultrasound to predict distal embolization for ST-segment elevation myocardial infarction. J Am Coll Cardiol. 2007;50:1641–1646. DOI: 10.1016/j.jacc.2007.06.051

[67] Claessen BE, Maehara A, Fahi M, Xu K, Stone GW, Mintz GS. Plaque composition by intravascular ultrasound and distal embolization after percutaneous coronary intervention. JACC Cardiovasc Imaging. 2012;S111–S118. DOI: 10.1016/j.jcmg.2011.11.018

[68] Jang JS, Jin HY, Seo JS, Yang TH, Kim DK, Park YA, Cho KI, Park YH, Kim DS. Meta-analysis of plaque composition by intravascular ultrasound and its relation to distal embolization after percutaneous coronary intervention. Am J Cardiol. 2013;111:968–972. DOI: 10.1016/j.amjcard.2012.12.016

[69] Stone PH, Saito S, Takahashi S, Makita Y, Nakamura S, Kawasaki T, Takahashi A, Katsuki T, Nakamura S, Namiki A, Hirohata A, Matsumura T, Yamazaki S, Yokoi H, Tanaka S, Otsuji S, Yoshimachi F, Honye J, Harwood D, Reitman M, Coskun AU, Papafaklis MI, Feldman CL. Prediction of progression of coronary artery disease and clinical outcomes using vascular profiling of endothelial shear stress and arterial plaque characteristics: the PREDICTION study. Circulation. 2012;126:172–181. DOI: 10.1161/CIRCULATIONAHA.112.096438

[70] Cheng JM, Garcia-Garcia HM, de Boer SP, Kardys I, Heo JM, Akkerhuis KM, Oemrawsingh RM, van Domburg RT, Ligthart J, Witberg KT, Regar E, Serruy PW, van Geuns RJ, Boersma E. In vivo detection of high-risk coronary plaques by radiofrequency intravascular ultrasound and cardiovascular outcome: results of the ATHEROREMO-IVUS study. Eur Heart J. 2014;35:639–647. DOI: 10.1093/eurheartj/eht484

[71] Huang D, Swanson EA, Lin CP, Schuman JS, Stinson WG, Chang W, Hee MR, Flotte T, Gregory K, Puliafito CA, et al. Optical coherence tomography. Science. 1991;254:1178–1181. DOI: 10.1126/science.1957169

[72] Prati F, Regar E, Mintz GS, Arbustini E, Di Mario C, Jang IK, Akasaka T, Costa M, Guagliumi G, Grube E, Ozaki Y, Pinto F, Serruys PW. Expert review document on methodology, terminology, and clinical applications of optical coherence tomography: physical principles, methodology of image acquisition, and clinical application for assessment of coronary arteries and atherosclerosis. Eur Heart J. 2010;31:401–415. DOI: 101093/eurheartj/ehp433

[73] Ozaki Y, Okumura M, Ismail TF, Naruse H, Hattori K, Kan S, Ishikawa M, Kawai T, Takagi Y, Ishii J, Prati F, Serruys PW. The fate of incomplete stent apposition with drug-eluting stents: an optical coherence tomography-based natural history study. Eur Heart J. 2010;31:1470–1476. DOI: 10.1093/eurheartj/ehq066

[74] Onuma Y, Serruys PW, Perkins LE, Okamura T, Gonzalo N, Garcia-Garcia HM, Regar E, Kamberi M, Powers JC, Rapoza R, van Beusekom H, van der Giessen W, Virmani R. Intracoronary optical coherence tomography and histology at 1 month and 2, 3, and 4 years after implantation of everolimus-eluting bioresorbable vascular scaffolds in a porcine coronary artery model: an attempt to decipher the human optical coherence tomography images in the ABSORB trial. Circulation. 2010;122:2288–2300. DOI: 10.1161/CIRCULATIONAHA.109.921528

[75] Radu MD, Falk E. In search of vulnerable features of coronary plaques with optical coherence tomography: is it time to rethink the current methodological concepts? Eur Heart J. 2012;33:9–12. DOI:10.1093/eurheartj/ehr290

[76] Takarada S, Imanishi T, Liu Y, Ikejima H, Tsujioka H, Kuroi A, Ishibashi K, Komukai K, Tanimoto T, Ino Y, Kitabata H, Kubo T, Nakamura N, Hirata K, Tanaka A, Mizukoshi M, Akasaka T. Advantage of next-generation frequency-domain optical coherence tomography compared with conventional time-domain system in the assessment of coronary lesion. Catheter Cardiovasc Interv. 2010;75:202–206. DOI: 10.1002/ccd.22273

[77] Madder RD, Smith JL, Dixon SR, Goldstein JA. Composition of target lesions by near-infrared spectroscopy in patients with acute coronary syndrome versus stable angina. Circ Cardiovasc Interv. 2012;5:55–61. DOI: 10.1161/CIRCINTERVENTIONS.111.963934

[78] Brilakis ES, Banerjee S. How to detect and treat coronary fibroatheroma: the synergy between IVUS and NIRS. JACC Cardiovasc Imaging. 2015;8:195–197. DOI: 10.1016/j.jcmg.2014.11.009

[79] Puri R, Madder RD, Madden SP, Sum ST, Wolski K, Muller JE, Andrews J, King KL, Kiyoko K, Uno K, Kapadia SR, Tuzcu EM, Nissen SE, Virmani R, Maehara A, Mintz GS, Nicholls SJ. Near-infrared spectroscopy enhances intravascular ultrasound assessment of vulnerable coronary plaque. A combined pathological and in vivo study. Arterioscler Thromb Vasc Biol. 2015;35:2423–2431. DOI: 10.1161/ATVBAHA.115.306118

[80] Brugaletta S, Garcia-Garcia HM, Serruys PW, de Boer S, Ligthart J, Gomez-Lara J, Witberg K, Diletti R, Wykrzykowska J, van Geuns RJ, Schultz C, Regar E, Duckers HJ, van Mieghem N, de Jaegere P, Madden SP, Muller JE, van der Steen AF, van der Giessen WJ, Boersma E. NIRS and IVUS for characterization of atherosclerosis in patients undergoing coronary angiography. JACC Cardiovasc Imaging. 2011;4:647–655. DOI: 10.1016/j.jcmg.2011.03.013

[81] Pu J, Mintz GS, Brilakis ES, Banerjee S, Abdel-Karim A-RR, Maini B, Biro S, Lee JB, Stone GW, Weisz G, Maehara A. In vivo characterization of coronary plaques: novel findings from comparing greyscale and virtual histology intravascular ultrasound and near-infrared spectroscopy. Eur Heart J. 2012; 33:372–383. DOI: 10.1093/eurheartj/ehr387

[82] Yonetsu T, Suh W, Abtahian F, Kato K, Vergallo R, Kim SJ, Jia H, McNulty I, Lee H, Jang IK. Comparison of near-infrared spectroscopy and optical coherence tomography for detection of lipid. Catheter Cardiovasc Interv. 2014;84:710–717. DOI: 10.1002/ccd.25084

[83] Roleder T, Kovacic JC, Ali Z, Sharma R, Cristea E, Moreno P, Sharma SK, Narula J, Kini AS. Combined NIRS and IVUS imaging detects vulnerable plaque using a single catheter system: a head-to-head comparison with OCT. EuroIntervention. 2014;10:303–311. DOI: 10.4244/EIJV1013A53

[84] Fur E, Brilakis ES. Comparative intravascular imaging for lipid core plaque: VH-IVUS vs OCT vs NIRS. J Invasive Cardiol. 2013;25:9A–13A. ISSN: 1557-2501

[85] Bourantas CV, Garcia-Garcia HM, Naka KK, Sakellarios A, Athanasiou L, Fotiadis DI, Michalis LK, Serruys PW. Hybrid intravascular imaging, current applications and prospective potential in the study of coronary atherosclerosis. J Am Coll Cardiol. 2013;61:1369–1378. DOI: 10.1016/j.jacc.2012.10.057

[86] Madder RD, Steinberg DH, Anderson D. Multimodality direct coronary imaging with combined near-infrared spectroscopy and intravascular ultrasound: initial US experience. Catheter Cardiovasc Interv. 2013;81:551–557. DOI: 10.1002/ccd.23358

[87] Madder RD, Wohns DH, Muller JE.Detection by intracoronary near-infrared spectroscopy of lipid core plaque at culprit sites in survivors of cardiac arrest. J Invasive Cardiol. 2014;26:78–79. DOI:

[88] Gebhard C, L'Allier PL, Tardif JC. Near-infrared spectroscopy for cardiovascular risk assessment? Not ready for primetime. Eur Heart J. 2014;35:263–265. DOI: 10.1093/eurheartj/eht361

[89] Narula J, Nakano M, Virmani R, Kolodgie FD, Petersen R, Newcomb R, Malik S, Fuster V, Finn AV. Histopathologic characteristics of atherosclerotic coronary disease and implications of the findings for the invasive and noninvasive detection of vulnerable plaques. J Am Coll Cardiol. 2013; 61:1041–1051. DOI: 10.1016/j.jacc.2012.10.054

[90] Patel D, Hamamdzic D, Llano R, Patel D, Cheng L, Fenning RS, Bannan K, Wilensky RL. Subsequent development of fibroatheromas with inflamed fibrous caps can be predicted by intracoronary near infrared spectroscopy. Arterioscler Thromb Vasc Biol. 2013; 33:347–353. DOI: 10.1016/ATVBAHA.112.300710

[91] Kang SJ, Mintz GS, Pu J, Sum ST, Madden SP, Burke AP, Xu K, Goldstein JA, Stone GW, Muller JE, Virmani R, Maehara A. JACC Cardiovasc Imaging. 2015;8:184–194. DOI: 10.1016/j.jcmg.2014.09.021

[92] Oemrawsingh RM, Cheng JM, García-García HM, van Geuns R-J, de Boer SPM, Simsek C, Kardys I, Lenzen MJ, van Domburg RT, Regar E, Serruys PW, Akkerhuis KM, Boersma E. Near-infrared spectroscopy predicts cardiovascular outcome in patients with coronary artery disease. J Am Coll Cardiol. 2014; 64:2510–2518. DOI: 10.1016/j.jacc.2014.07.998

[93] Madder RD, Husaini M, Davis AT, VanOosterhout S, Kan M, Wohns D, McNamara RF, Wolschleger K, Gribar J, Collins JS, Jacoby M, Decker JM, Hendricks M, Sum ST, Madden

S, Ware JH, Muller JE. Large lipid-rich coronary plaques detected by near-infrared spectroscopy at non-stented sites in the target artery identify patients likely to experience future major adverse cardiovascular events. Eur Heart J. 2016;17:393–399.

[94] Witzenbichler B, Maehara A, Weisz G, Neumann FJ, Rinaldi MJ, Metzger DC, Henry TD, Cox DA, Duffy PL, Brodie BR, Stuckey TD, Mazzaferri EL, Xu K, Parise H, Mehran R, Mintz GS, Stone GW. Relationship between intravascular ultrasound guidance and clinical outcomes after drug-eluting stents: the assessment of dual antiplatelet therapy with drug-eluting stents (ADAPT-DES) study. Circulation. 2014;129:463–470. DOI : 10.1161/CIRCULATIONAHA.113.003942

[95] Jang JS, Song YJ, Kang W, Jin HY, Seo JS, Yang TH, Kim DK, Cho KI, Kim BH, Park YH, Je HG, Kim DS. Intravascular ultrasound-guided implantation of drug-eluting stents to improve outcome: a meta-analysis. JACC Cardiovasc Interv. 2014;7:233–243. DOI: 10.1016/j.jcin.2013.09.013

[96] Ahn JM, Kang SJ, Yoon SH, Park HW, Kang SM, Lee JY, Lee SW, Kim YH, Lee CW, Park SW, Mintz GS, Park SJ. Meta-analysis of outcomes after intravascular ultrasound-guided versus angiography-guided drug-eluting stent implantation in 26,503 patients enrolled in three randomized trials and 14 observational studies. Am J Cardiol. 2012;109:60–66. DOI: 10.1016/j.amjcard.2013.12.043

[97] Dixon SR, Grines CL, Munir A, Madder RD, Safian RD, Hanzel GS, Pica MC, Goldstein JA. Analysis of target lesion length before coronary artery stenting using angiography and near-infrared spectroscopy versus angiography alone. Am J Cardiol. 2012;109:60–66. DOI: 10.1016/j.amjcard.2011.07.068

[98] Hanson ID, Goldstein JA, Dixon SR, Stone GW. Comparison of coronary artery lesion length by NIRS-IVUS versus angiography alone. Coron Artery Dis. 2015;26:484–489. DOI: 10.1097/MCA.0000000000000263

[99] Saeed B, Banerjee S, Brilakis ES. Slow flow after stenting of a coronary lesion with a large lipid core plaque detected by near-infrared spectroscopy. EuroIntervention. 2010;6:545.

[100] Awata M, Kotani J, Uematsu M, Morozumi T, Watanabe T, Onishi T, Iida O, Sera F, Nanto S, Hori M, Nagata S. Serial angiographic evidence of incomplete neointimal coverage after sirolimus-eluting stent implantation: comparison with bare-metal stents. Circulation. 2007;116:910–916. DOI: 10.1161/CIRCULATIONAHA.105.6609057

[101] Waxman S, Freilich MI, Stuer MJ, Shishkov M, Bilazarian S, Virmani R, Bouma BE, Tearney GJ. A case of lipid core plaque progression and rupture at the edge of a coronary stent: elucidating the mechanisms of drug-eluting stent failure. Circ Cardiovasc Interv. 2010;3:193–196. DOI: 10.1161/CIRCINTERVENTIONS.109.917955

[102] Stouffer GA.The use of near-infrared spectroscopy to optimize stent length.J Invasive Cardiol. 2013;25:5A–8A. ISSN: 1557-2501

[103] Joner M, Finn AV, Farb A, Mont EK, Kolodgie FD, Ladich E, Kutys R, Skorija K, Gold HK, Virmani R. Pathology of drug-eluting stents in humans: delayed healing and late thrombotic risk. J Am Coll Cardiol. 2006;48:193–202. DOI: 10.1016/j.jacc.2006.03.042

[104] Oyabu J, Ueda Y, Ogasawara N, Okada K, Hirayama, Kodama K. Angioscopic evaluation of neointima coverage: sirolimus drug-eluting stent versus bare metal stent. Am Heart J. 2006;152:1168–1174. DOI: 10.1016/j.ahj.2006.07.025

[105] Finn AV, Nakazawa G, Ladich E, Kolodgie FD, Virmani R. Does underlying plaque morphology play a role in vessel healing after drug-eluting stent implantation. JACC Cardiovasc Imaging. 2008;1:1485–1488. DOI: 10.1016/j.jcmg.2008.04.007

[106] Nakazawa G, Finn AV, Joner M, Ladich E, Kutys R, Mont EK, Gold HK, Burke AP, Kolodgie FD, Virmani R. Delayed arterial healing and increased late stent thrombosis at culprit sites after drug-eluting stent placement for acute myocardial infarction patients: an autopsy study. Circulation. 2008;118:1138–1145. DOI: 10.1161/CIRCULATIONAHA.107.762047

[107] Otsuka F, Byrne RA, Yahagi K, Mori H, Ladich E, Fowler DR, Kutys R, Xhepa E, Kastrati A, Virmani R, Joner M. Neoatherosclerosis: overview of histopathologic findings and implications for intravascular imaging assessment. Eur Heart J. 2015;36:2147–2159. DOI: 10.1093/eurheartj/ehv205

[108] Ali ZA, Roleder T, Narula J, Mohanty BD, Baber U, Kovacic JC, et al. Increased thin-cap neoatheroma and periprocedural myocardial infarction in drug-eluting stent restenosis: multimodality intravascular imaging of drug-eluting and bare-metal stents. Circ Cardiovasc Interv. 2013; 6:507–517. DOI: 10.1161/CIRCINTERVENTIONS.112.000248

[109] Madder RD, Khan M, Husaini M, Chi M, Dionne S, VanOosterhout S, Borgman A, Collins JS, Jacoby M. Combined near-infrared spectroscopy and intravascular ultrasound imaging of pre-existing coronary artery stents. Can near-infrared spectroscopy reliably detect neoatherosclerosis? Circ Cardiovasc Imaging. 2016;9:e003576. DOI: 10.1161/CIRCIMAGING.115.003576

[110] Ramcharitar S, Gonzalo N, van Geuns RJ, Garcia-Garcia HM, Wykrzykowska JJ, Ligthart JM. First case of stenting of a vulnerable plaque in the SECRIIT I trial-the dawn of a new era? Nat Rev Cardiol. 2009;6:374–378. DOI: 10.1038/nrcardio.2009.34

[111] Finn AV, Joner M, Nakazawa G, Kolodgie F, Newell J, John MC, et al. Pathological correlates of late drug-eluting stent thrombosis: strut coverage as a marker of endothelialization. Circulation. 2007;115:2435–2441. DOI: 10.1161/CIRCULATIONAHA.107.693739

[112] Brugaletta S, Radu MD, Garcia-Garcia HM, Heo JH, Farooq V, Girasis C, et al. Circumferential evaluation of the neointima by optical coherence tomography after ABSORB bioresorbable vascular scaffold implantation: can the scaffold cap the plaque? Atherosclerosis. 2012;221:106–112. DOI: 10.1016/j.atherosclerosis.2011.12.008

[113] Heusch G, Kleinbongard P, Böse D, Levkau B, Haude M, Schulz R, Erbel R. Coronary microembolization: From bedside to bench and back to bedside. Circulation. 2009;120:1822–1836. DOI:10.1161/CIRCULATIONAHA.109.888784

[114] Prasad A, Singh M, Lerman A, Lennon RJ, Holmes DR Jr, Rihal CS. Isolated elevation in troponin T after percutaneous coronary intervention is associated with higher long-term mortality. J Am Coll Cardiol. 2006;48:1765–1770. DOI: 10.1016/j.jacc.2006.04.102

[115] Tanaka A, Kawarabayashi T, Nishibori Y, Sano T, Nishida Y, Fukuda D, Shimada K, Yoshikawa J. No-reflow phenomenon and lesion morphology in patients with acute myocardial infarction. Circulation. 2002;105:2148–2152. DOI: 10.1016/01.CIR.0000015697.59592.07

[116] Limbruno U, De Carolo M, Pistolesi S, Micheli A, Petronio AS, Camacci T, Fontanini G, Balbarini A, Mariani M, De Caterina R. Distal embolization during primary angioplasty: histopathologic features and predictability. Am Heart J. 2005;150:102–108. DOI: 10.1016/j.ahj.2005.01.016

[117] Kotani J, Nanto S, Mintz GS, Kitakaze M, Ohara T, Morozumi T, Nagata S, Hori M. Plaque gruel of atheromatous coronary lesion may contribute to the no-reflow phenomenon in patients with acute coronary syndrome. Circulation. 2002;106:1672–1677. DOI: 10.1161/01.CIR.0000030189.27175.4E

[118] Kawamoto T, Okura H, Koyama Y, Toda I, Taguchi H, Tamita K, Yamamuro A, Yoshimura Y, Neishi Y, Toyota E, Yoshida K. The relationship between coronary plaque characteristics and small embolic particles during coronary stent implantation. J Am Coll Cardiol. 2007;50:1635–1640. DOI: 10.1016/j.jacc.2007.05.050

[119] Goldstein JA, Grines C, Fischell T, Virmani R, Rizik D, Muller J, Dixon SR. Coronary embolization following balloon dilatation of lipid-core plaques. JACC Cardiovasc Imaging. 2009;2:1420–1424. DOI: 10.1016/j.jcmg.2009.10.003

[120] Papayannis AC, Abdel-Karim A-RR, Mahmood A, Rangan B V, Makke LB, Banerjee S, et al. Association of coronary lipid core plaque with intrastent thrombus formation: a near-infrared spectroscopy and optical coherence tomography study. Catheter Cardiovasc Interv. 2013; 81:488–493. DOI: 10.1002/ccd.23389

[121] Schultz CJ, Serruys PW, van der Ent M, Ligthart J, Mastik F, Garg S, et al. First-in-man clinical use of combined near-infrared spectroscopy and intravascular ultrasound: a potential key to predict distal embolization and no-reflow? J Am Coll Cardiol. 2010; 56:314. DOI: 10.1016/j.jacc.2009.10.090

[122] Garcia BA, Wood F, Cipher D, Banerjee S, Brilakis ES. Reproductibility of near-infrared spectroscopy for the detection of lipid core coronary plaques and observed changes after coronary stent implantation. Catheter Cardiovasc Interv. 2010;76:359–365. DOI: 10.1002/ccd.22500

[123] Brilakis ES, Abdel-Karim A-RR, Papayannis AC, Michael TT, Rangan B V, Johnson JL, et al. Embolic protection device utilization during stenting of native coronary artery lesions with large lipid core plaques as detected by near-infrared spectroscopy. Catheter Cardiovasc Interv. 2012; 80:1157–1162. DOI: 10.1002/ccd.23507

[124] Maini A, Buyantseva L, Maini B. In vivo lipid core plaque modification with percutaneous coronary revascularization: a near-infrared spectroscopy study. J Invasive Cardiol. 2013;25:293–295. DOI: PMID: 23735355

[125] Stone GW, Maehara A, Muller JE, Rizik DG, Shunk KA, Ben-Yehuda O, Généreux P, Dressler O, Parvataneni R, Madden S, Shah P, Brilakis ES, Kini AS. Plaque

characterization to inform the prediction and prevention of periprocedural myocardial infarction during percutaneous coronary intervention: the CANARY Trial (Coronary Assessment by Near-infrared of Atherosclerotic Rupture-prone Yellow). JACC Cardiovasc Interv. 2015;8:927–936. DOI: 10.1016/j.jcin.2015.01.032

[126] Vlaar PJ, Svilaas T, van der Horst IC, Diercks GF, Fokkema ML, de Smet BJ, van den Heuvel AF, Anthonio RL, Jessurum GA, Tan ES, Suurmeijer AJ, Zijlstra F. Cardiac death and reinfarction after 1 year in the thrombus aspiration during percutaneous coronary intervention in acute myocardial infarction study (TAPAS): a 1-year follow-up study. Lancet. 2008;371:1915–1920. DOI: 10.1016/S0140-6736(08)60833-8

[127] Frobert O, Lagerqvist B, Olivercrona GK, Omerovic E, Gudnason T, Maeng M, Aasa M, Angeras O, Calais F, Danielewicz M, Erlinge D, Hellsten L, Jensen U, Johansson AC, Karegren A, Nilsson J, Robertson L, Sandhall L, Sjögren I, Ostlund O, Harnek J, James SK. Thrombus aspiration during ST-segment elevation myocardial infarction. N Engl J Med. 2013;369:1587–1597. DOI: 10.1056/NEJMoa1308789

[128] Jolly SS, Cairns JA, Yusuf S, Meeks B, Pogue J, Rokoss MJ, Kedev S, Thabane L, Stankovic G, Moreno R, Gershlick A, Chowdhary C, Lavi S, Niemelä K, Steg PG, Bernat I, Xu Y, Cantor WJ, Overgaard CB, Naber CK, Cheema AN, Welsh RC, Bertrand OF, Avezum A, Bhindi R, Pancholy S, Rao SV, Natarajan K, ten Berg JM, Shestakovska O, Gao P, Widimsky P, Dzavik V. Randomized trial of primary PCI with or without routine manual thrombectomy. N Engl J Med. 2015;372:1389–1398. DOI: 10.1056/NEJMoa1415098

[129] Erlinge D, Harnek J, Goncalves I, Gotberg M, Muller JE, Madder RD. Coronary liposuction during percutaneous coronary intervention: evidence by near-infrared spectroscopy that aspiration reduces culprit lesion lipid content prior to stent placement. Eur Heart J Cardiovasc Imaging. 2015;16:316–324. DOI: 10.1093/ehjci/jeu180

[130] Simsek C, van Geuns RJ, Magro M, Boersma E, Garcia-Garcia HM, Serruys PW. Change in near-infrared spectroscopy of a coronary artery after 1-year treatment with high dose rosuvastatin. Int J Cardiol. 2012;157:e54–e56. DOI: 10.1016/j.ijcard.2011.09.047

[131] Simsek C, Garcia-Garcia HM, van Geuns RJ, Magro M, Girasis C, van Mieghem N, Lenzen M, de Boer S, Regar E, van der Giessen W, Raichlen J, Duckers HJ, Zijlstra F, van der Steen T, Boersma E, Serruys PW. The ability of high dose rosuvastatin to improve plaque composition in non-intervened coronary arteries: rationale and design of the integrated biomarker and imaging study-3 (IBIS-3). EuroIntervention. 2012;8:234–241. DOI: 10.4244/EIJV912A37

[132] Jang IK.Near infrared spectroscopy.Another toy or indispensable diagnostic tool? Circ Cardiovasc Interv. 2012;5:10–11. DOI:10.1161/CIRCINTERVENTIONS.111.967935

[133] Kaul S, Narula J. In search of the vulnerable plaque: is there any light at the end of the catheter? J Am Coll Cardiol. 2014; 64:2519–24.10.1016/j.jacc.2014.10.017

[134] Jang J-S, Song Y-J, Kang W, Jin H-Y, Seo J-S, Yang T-H, et al. Intravascular ultrasound-guided implantation of drug-eluting stents to improve outcome: a meta-analysis. J Am Coll Cardiol Cardiovasc Interv. 2014; 7:233–243. DOI: 10.1016/j.jcin.2013.09.013

Chapter 3

Highly Sensitive Singlet Oxygen Spectroscopic System Using InGaAs PIN Photodiode

Iwao Mizumoto, Hiroshi Oguma and Yostumi Yoshi

Additional information is available at the end of the chapter

http://dx.doi.org/10.5772/66644

Abstract

The spectrum of $^1O_2$ was measured by the InGaAs photodiode for an optical communication system with charge integration amplifier (InGaAs-CIA). The photo-excited current is charged in photodiode junction capacitance itself. The current is changed to the voltage about 10^{12} times without feedback resistance. The minimum detectable power of InGaAs CIA system with liquid nitrogen was achieved 0.1 fW of 10 sec integration time at the wavelength of 1.28 μm. The optical band pass filter-based system for ultra–low-level light detection was succeeded in spectrum measurement of $^1O_2$ by 13-LOOH with cytochrome c. The 8 channel InGaAs-CIA array system enables to achieve optical multichannel detection for ultra-low level light at 10^{-13} W from 10^{-15} W level in the near-infrared region. The optical resolution was about 200 nm by 1 channel. The spectrum of $^1O_2$ by mixing NaOCl and H_2O_2 was demonstrated. The shape of spectrum by $^1O_2$ was matched to that of measured by the spectrometer. The system was succeeded in instantaneous $^1O_2$ spectrum measurement without moving the wavelength dispersion device. The generation of $^1O_2$ by photo-excited Rose Bengal was fabricated to develop food antioxidant chemistry or source reagent of cosmetic product. The system uses super luminosity LED for excitation light source and InGaAs CIA. The $^1O_2$ generation will be controlled by the InGaAs-CIA monitoring system. The system will be used in the chemical plant of primary material production.

Keywords: InGaAs PIN photodiode, low-level light, singlet oxygen, charge integrating amplifier, spectroscopic system, multichannel, filter based, Rose Bengal

1. Introduction

Low-level light is emitted from various kinds of faint sources in the visible region. It is difficult to specify the emitting source chemically because the spectra overlap each other. In the near infrared (NIR), existence of the substance of biological material for emitting is not scarce,. Also the thermal back ground noise is exceedingly released compared with the middle-infrared region or the far-infrared region. The problem of measuring exceedingly low levels light from various kinds of faint sources is of considerable interest and importance. Especially, $^1O_2$ is emitted with low-level light at the 1.27 μm of NIR optical band. $^1O_2$ is one of active oxygen species from the biological material. The chemiluminescent substance is scarce at the NIR region, the use of spectroscopic $^1O_2$ emission will be available as a chemical and physical analytical tool in. There are many biochemists, pathologists, and agricultural chemists who focus on such a weak optical signal in the NIR [1]. Since silicon photodiode have no optical sensitivity in the NIR, Ge PIN photodiodes with phase sensitive amplifier are used in NIR [1]. Johnson noise is given by the equation, Eq. (1),

$$I_j = \sqrt{\frac{4\,k_D\,T_n}{R_{sh}}} \tag{1}$$

where I_j is thermal noise current, k_D is Boltzmann constant, T_n is temperature in kelvin, R_{sh} is shunt resistance in photodiode. R_{sh} is in inverse proportion to dark current. The shot noise (dark) is given by the next equation, Eq. (2),

$$I_{sD} = \sqrt{2e\,I_{dn}} \tag{2}$$

where I_{sD} is shot noise in darkness, e is the electron charge, I_{dn} is the dark current. At the operating temperature of 77 K, the device thermal noise and shot noise are not dominant [2, 3, 4]. The optical sensitivity is mainly decided by the dark current. The principal noise current of such a detector with a transimpedance amplifier (TIA) is given by the equation, Eq. (3),

$$I_n = \sqrt{I_{dn}\,e\Delta f} \tag{3}$$

where I_n is noise current, I_{dn} is dark current, e is the electron charge, and Δf is the bandwidth. **Figure 1** shows the temperature dependence of the dark current InGaAs photodiode (Fujitsu FID13Y23WY) and Ge PIN photodiode (Fujitsu FID13R53WZ) for the optical communication system. The dark current of an InGaAs photodiode at 77 K is three orders of magnitude less than of a Ge PIN photodiode.

Figure 2 shows temperature dependence of quantum efficiency the InGaAs photodiode and the Ge PIN photodiode. The quantum efficiency of the Ge PIN photodiode decreases from the temperature of liquid nitrogen. That of the InGaAs maintains to the temperature of liquid helium. The dark current and quantum efficiency of the Ge PIN photodiode decrease in

http://dx.doi.org/10.5772/66644

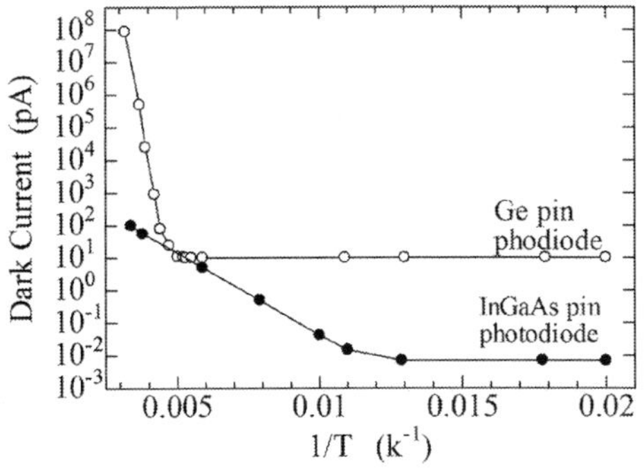

Figure 1. Temperature dependence of dark current InGaAs photodiode and Ge pin photodiode (identical to the one published in authors' previous work [5]).

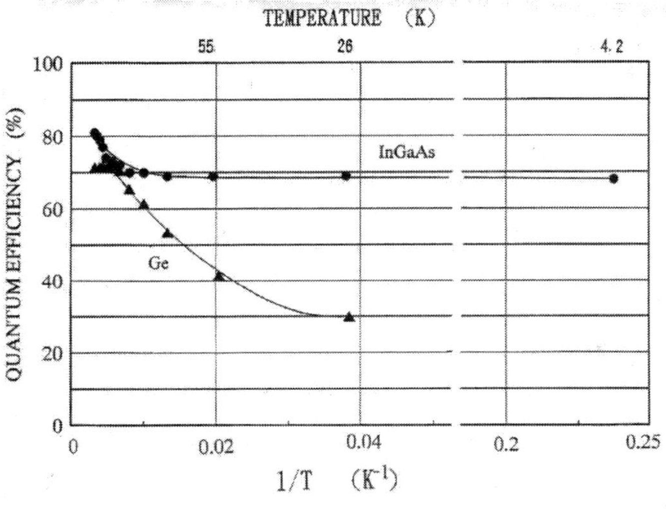

Figure 2. Temperature dependence of quantum efficiency InGaAs photodiode and Ge pin photodiode.

proportion to the device temperature. The device temperature of the InGaAs PIN photodiode has little influence to the quantum efficiency. Therefore, the InGaAs PIN photodiode is more suitable than the Ge PIN photodiode for detecting low-level light in the NIR region.

2. Circuit of detection system

The impedance of an InGaAs PIN photodiode cooled to 77 K is so high (100 TΩ). It can be operated with a charge integrating amplifier (CIA) [2, 4, 5]. The amount accumulated charge

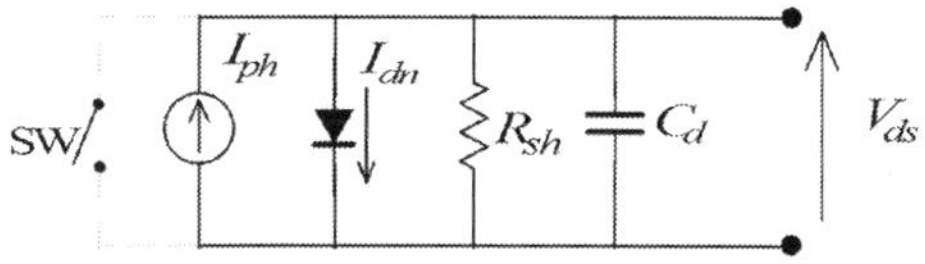

Figure 3. Diode equivalent circuit (identical to the one published in authors' previous work [5]).

in this configuration is directly measured as voltage. The photocurrent or carrier in the CIA is converted to voltage through the capacitance of a photodiode. **Figure 3** shows the diode equivalent circuit with a simplified signal source. R_{sh} is so high and I_{dn} is very small at the temperature 77 K.

The CIA output voltage is given by the next equation, Eq. (4).

$$V_{ds} = \frac{\int_0^{\tau} I_{ph}\, dt}{C_d} \quad (4)$$

Where V_{ds} is the voltage of the signal output. I_{ph} is the photocurrent, τ is integration time, C_d is the capacitance of the detector. The InGaAs-CIA enables photocurrent to the voltage without a feedback register. The minimum detectable incident power is given by

$$P_{min} = \frac{h\nu}{\eta}\left[\frac{e_{am}\sqrt{\Delta f}}{e\tau} C_d + \sqrt{\frac{I_{dn}}{2e\tau}}\right] \quad (5)$$

Where P_{min} is the minimum detectable power, I_{dn} is the dark current, e is the electron charge η is the quantum efficiency of the photodiode, Δf is the bandwidth, e_{am} is the noise voltage of the amplifier, τ is the integration time, and C_d is the capacitance [5]. At the condition $I_{dn} = 5 \times 10^{-15}$ A, $\eta = 0.7$, $C_d = 30$ pF, $e_{am} = 100$ nVHz$^{-1/2}$, $\tau = 10$ sec, $\Delta f = 100$ Hz, the calculated P_{min} value is 6×10^{-17} W.

Figure 4 shows the InGaAs-CIA detailed circuit diagram. The InGaAs PIN photodiode (Fujitsu FID13Y23WY) was used for CIA. The output voltage measured an amount of photocurrent. A dual n-channel J. FET (2N6483) differential amplifier minimized current drift error from fluctuation of the temperature. The FET source follower circuit reduces the output impedance for reduction of inductive noise. The timer IC (NE555) with p-channel MOSFET (3SJ11A) controls the time of charge accumulating.

Figure 5 shows the chart records of the InGaAs-CIA. The integration time of 10 sec yields differential voltage of 130 mV between on-emission light of 10^{-15} W and off light. The voltage fluctuation in 10 sec was 10 mV. The minimum detectable power was measured by the LED optical source (ADVANTEST TQ-28 at 1.28 μm with FWHM 30 nm) and attenuates NIR ND filters. The system obtained minimum detectable optical power 10^{-16} W at 1.28 μm with 10 sec. The result corresponded to the predicted value of equation.

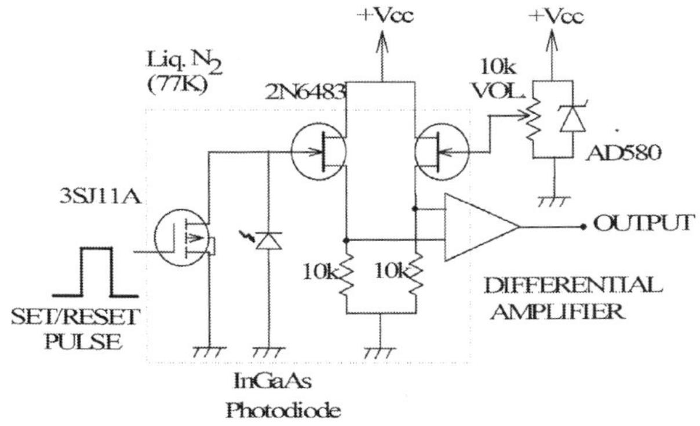

Figure 4. Circuit diagram of the InGaAs-CIA (almost identical to the one published in authors' previous work [5, 6]).

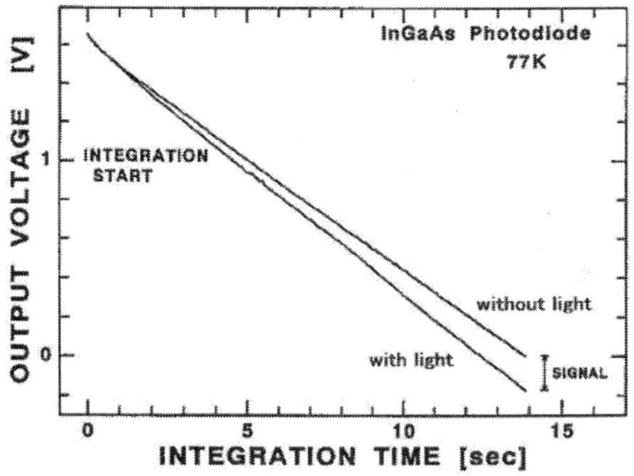

Figure 5. Chart records of the InGaAs-CIA (identical to the one published in authors' previous work [5]).

Figure 6 shows the theoretical and experimental results of the minimum detectable power. The solid and dashed lines show the experimental and theoretical results, respectively. The minimum detectable power of 10^{-16} W was achieved at 1.28 μm wavelength.

At the theoretical line above 10 sec, the measured minimum detectable power was saturated. The influence of electrical FET device noise, 1/*f* low frequency noise, the saturation capacity of the dark current, and leak photocurrent are considered.

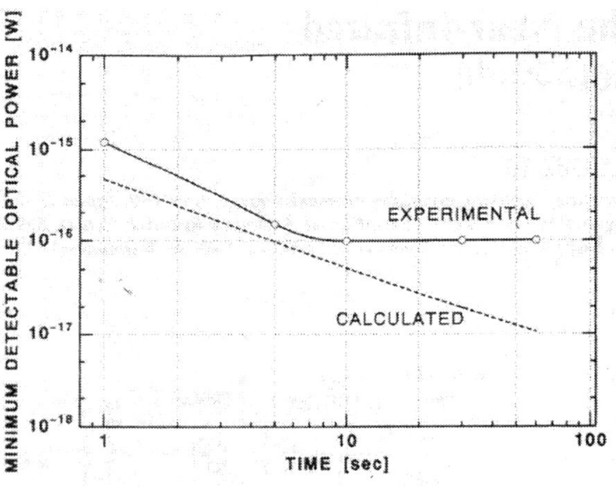

Figure 6. Minimum detectable power dependence of integration time (InGaAs pin photodiode at 77 K) (Identical to the one published in authors previous work [5]).

3. Application system of spectroscopic measurements in NIR

3.1. Filter-based high efficiency spectroscopic system

In the NIR region, low-level light has important information of biological, biochemistry, agricultural chemistry, and photochemistry. The emission phenomenon or spectrum is expected to explain the emission mechanisms. The biophoton or biological emission has extremely low-level light. The spectrum of the emission has broadly wavelength. The change of emission intensity is at a slow speed. An InGaAs-CIA is effective to accumulate excited photocurrent of the emission. It is difficult to focus for the spectrometer by lenses as the solid angle of biophoton or chemiluminescence from the living organism substance has a great angle of radiation and incoherent characteristic. Normally, the transmission of the spectrometer is less than 10%. A great solid angle as possible and short distance between optical source and dispersive element is important for the effective measurement. From these points of view, a high-efficiency filter-based InGaAs-CIA spectroscopic system was fabricated [6]. **Figure 7** shows a schematic diagram of the system. Fifteen interference filters (vacuum optics corporation of Japan) was used of the spectroscopic system. The measurement spectra range of the system has 1.0– 1.6 μm with 30–35 nm optical resolution. The interference filters have the average transmission of 82 ± 6%.

We measured the emission spectrum of $^1O_2$ at 1.27 μm to test the performance for our filter-based spectrometer. The emission with $^1O_2$ is very weak because it derives from a forbidden transition. The spectrum of $^1O_2$ produced by 13-LOOH (2 mM) with cytochrome c (10 μM), NH_3-NH_4Cl (0.04 M), D_2O is shown in **Figure 8**. We have observed a derivation biochemistry emission spectrum of $^1O_2$ produced by this reaction. It means that $^1O_2$ generated by oxidized 13-LOOH with cytochrome c of protein material of human body. The reaction is known as Russell's mechanism [7, 8].

Figure 9 shows Russell's mechanism. The chemical equation is shown in Eq. (6).

http://dx.doi.org/10.5772/66644

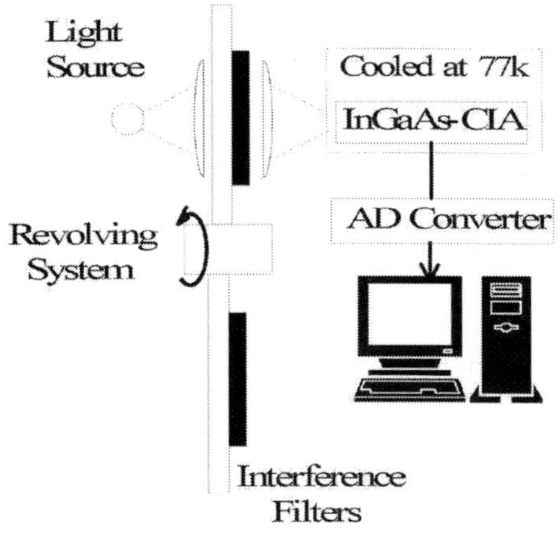

Figure 7. Interference filter based spectroscopic system (identical to the one published in authors' previous work [6]).

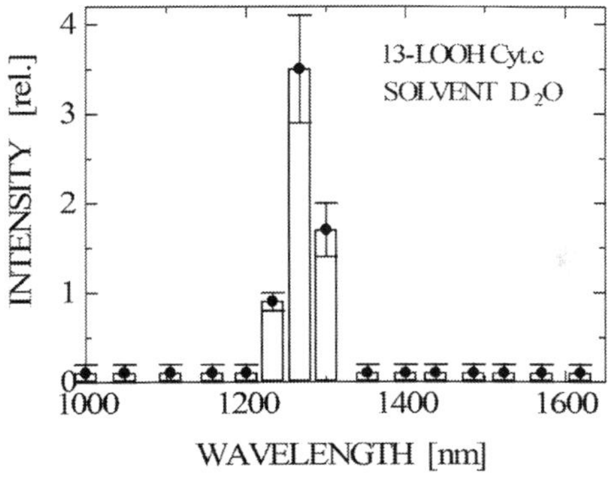

Figure 8. $^1O_2$ spectral of 13-LOOH with cytochrome c, D_2O (identical to the one published in authors' previous work [6]).

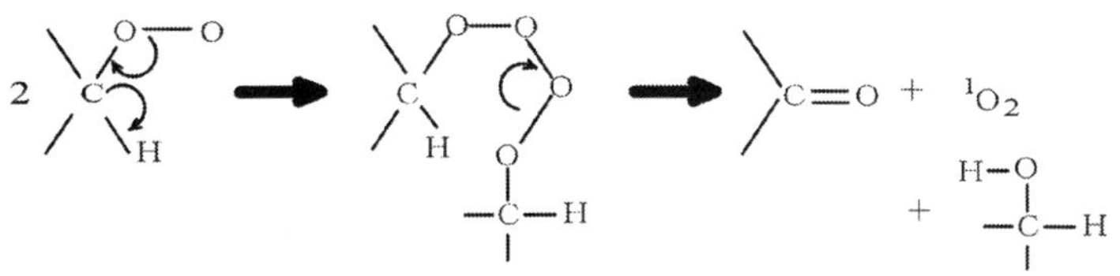

Figure 9. The Russell's mechanism [7].

$$ROO\cdot + ROO\cdot \rightarrow R=O + ROH + {}^1O_2 \quad (6)$$

The photo-generation in the visible region has been derived from $^1O_2$ in the past because the lipid peroxidation product is proportion to the emission intensity in the visible region. However, the emission by excited carbonyl involves in Russell's mechanism. Tyrosine and tryptophan emits at the wavelength of 500– 600 nm [7, 8]. A traditional method to detect $^1O_2$-involved chemical technique has no specificity for $^1O_2$ detection. The sensitivity of $^1O_2$ detection in near infrared enables to detect $^1O_2$ specificity because there is no emission material excepted. The heme compounds in the living organism include hemoglobin (blood), myoglobin (muscle), and cytochrome c (mitochondria). These heme compounds play important role in living body. In the case of inflammation in biological membranes, $^1O_2$ is generated by Russell's mechanism. The heme compounds perform catalysis on this occasion. We titrate the generation of $^1O_2$ by adding NaOCl after the excess H_2O_2 was put in the reaction chamber. The optical intensity of $^1O_2$ by 13-LOOH with cytochrome c, in D_2O in near infrared was compared with that of a typical $^1O_2$ generating method by mixed NaOCl and H_2O_2. The optical yield of $^1O_2$ by 13-LOOH with cytochrome c, in D_2O was decided using a calibration curve. **Figure 10** shows the calibration curve of the $^1O_2$ optical yield at the wavelength of 1.27 μm using the NaOCl-H_2O_2 system.

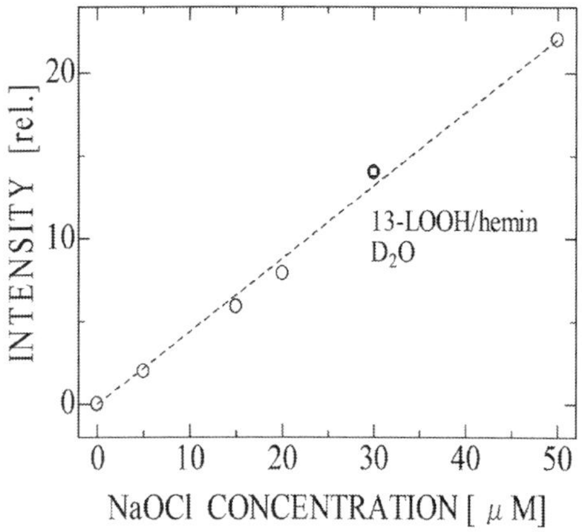

Figure 10. Calibration curve of the $^1O_2$ optical yield at the wavelength of 1.27 μm using NaOCl-H_2O_2 system.

These results show the degree of risk in case of $^1O_2$ generated from the heme compound. The InGaAs-CIA filter based spectroscopic system will be expected as an analysis equipment of oxidant stress by $^1O_2$.

3.2. Multichannel spectroscopic system

The multichannel spectroscopic system is available for measurement of fast emitting phenomena because the optical dispersion device of spectroscopic cannot be avoided to moving for wavelength shifting. In the visible region, silicon CCD camera with photoelectron multiplication function for detecting low-level light is commercially available. A silicon multichannel photo device at the wavelength of 1 μm has no photosensitivity. The light detecting materials for the NIR is easily influenced dark current. The commercially available InGaAs CCD or NIR photomultiplier of InGaAs photocathode material is very expensive. Additionally, this system is short of optical sensitivity for $^1O_2$ measurement in the NIR. A highly sensitive 8 channel InGAs-CIA spectroscopic system was developed for $^1O_2$ measurement. A commercially available InGaAs PIN photodiode for an optical communication device was used to stabilize parts supply. In case of employing the high sensitivity multichannel array, each photodiode or Avalanche photo diode device needs a lock-in amplifier with TIA circuits or photon counting equipment. Such a system is difficult to fabricate in reality because it needs many lock-in amplifier or photon counter in proportion to the number of channels. The InGaAs-CIA multichannel system has respectively signal transduction system to voltage from current. The system was allowed simple circuit, signal processing, and signal acquiring system with a low bit AD converter. The outline of InGaAs photodiode array is shown in **Figure 11**. The commercially available photodiode Fujitsu FID13Y13TX has a diameter of 1 mmφ. The 8 photodiodes was fabricated without spacing. The wide size is about 8 mm. Hamamatsu Photonics Co. assembled wire bonding during photodiodes device and packaging.

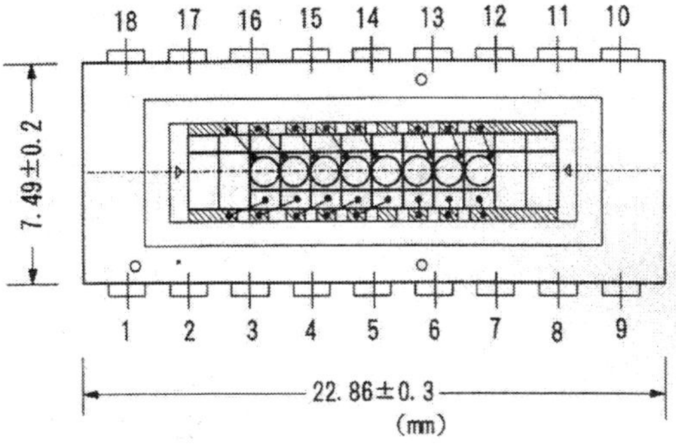

Figure 11. Outline of InGaAs photodiode array.

The block diagram of 8 channel multichannel InGaAs-CIA system is shown **Figure 12**. The fundamental circuit is as same as the monocyclic InGaAs-CIA system. The output signal of CIA array was connected to the AD converter for data acquisition after transmitted low impedance by J-FET source follower circuits. The system achieved minimum detectable optical power of 5×10^{-15}~10×10^{-15}. The deviation of sensitivity in each channel was inner single digit.

Figure 13 shows the block diagram of 8 array InGaAs-CIA multichannel spectroscopic system. The near-infrared light of the optical source was guided to dispersive element after collimator lens for parallel beam through an optical fiber of 100 μm core diameter. The NIR light was dispersed by the grating (Shimazu co. blaze wavelength of 1.2 μm, 300 line/1 mm, efficiency about 60%). The spectroscopic resolution of 1 channel was allowed about 75 nm

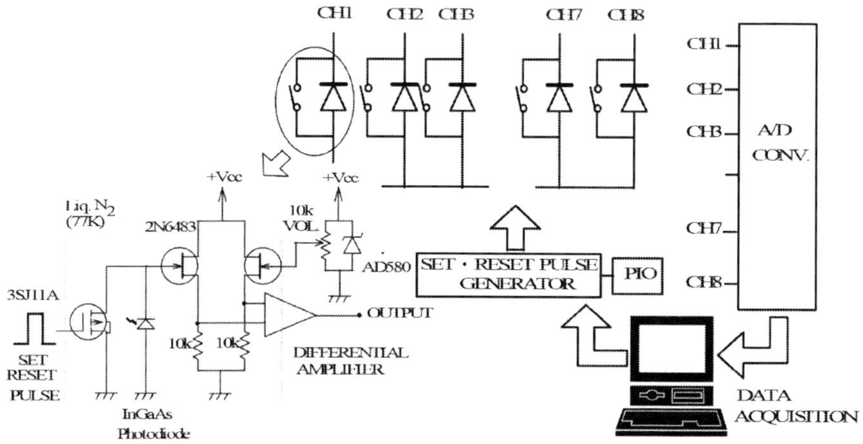

Figure 12. 8 multichannel InGaAs-CIA.

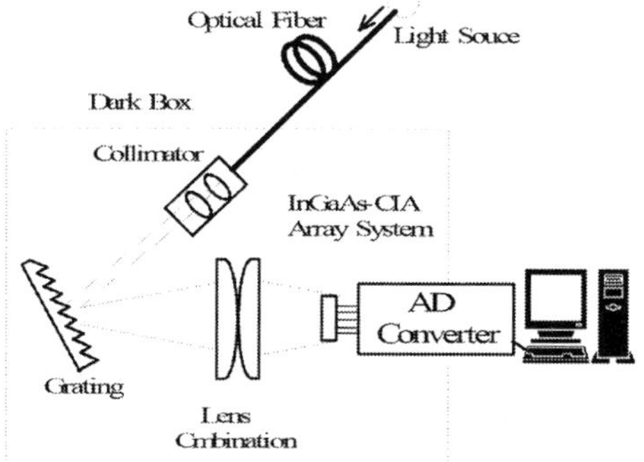

Figure 13. 8 array InGaAs-CIA spectroscopic system.

from 1000 to 1600 nm. The dispersion light was condensed on the photodiode array by a focal lens. The calibration of center wavelength on 8 photodiode was achieved using light from the monochromator as an optical band pass filter. The FWHM (full width at half maximum) $^1O_2$ spectrum was broadened about 100 nm at the center of 1265 nm. It needs no detailed resolution for spectrum measurement. In the NIR region, the chemical fluorescence of a biological material has no scarcely existence except $^1O_2$. The low-level light at the center wavelength of 1.27 μm enables to eliminate the thermal background noise.

A proof of any channel of photodiode array was carried out with the wavelength shift by a spectral apparatus of resolution 30 nm of the monochromator from 1000 to 1600 nm. **Figure 14** shows $^1O_2$ spectral of chemiluminescence with NaOCl mixed H_2O_2. The spectral of center wavelength was 1.28 μm with 50 nm of full width at half maximum. **Figure 14** shows the spectral of $^1O_2$.

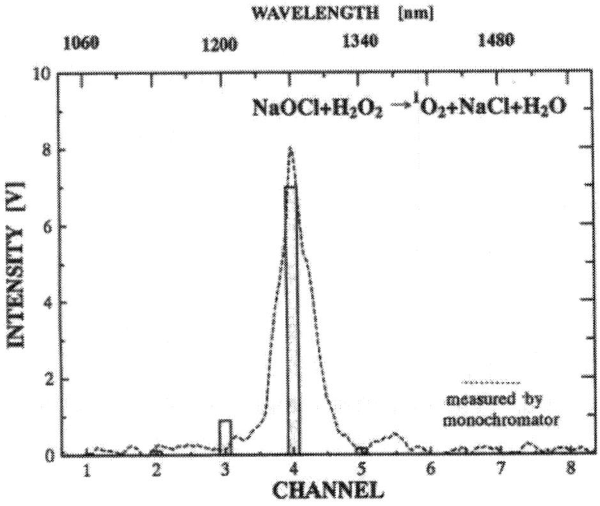

Figure 14. $^1O_2$ spectral of chemiluminescence.

The 8 multichannel spectroscopic system succeeded in measurement of $^1O_2$ spectral as same as measurement by the monochromatic spectrometer [9]. The system succeeded in simultaneous multiwavelength of $^1O_2$ spectrum measurement.

This measurement system will be useful for fast optical phenomenon, periodic emission, non-destructive measurement for melon, watermelon, and meat with extremely optical intensity damped in the NIR spectroscopy. The absorption band of protein, sugar, and lipid has a broad spectrum in the NIR. The detection sensitivity and measurement speed are needed. Our multichannel detection system is suitable for such a measurement condition.

3.3. $^1O_2$ monitoring system for antioxidant chemical test

In development of antioxidant of food chemistry, found metal material of beauty product, color, the super oxidation power allows the experiments of acid resistance and reaction promotion. In case of acid resistance test, the generation of $^1O_2$ from chemical reaction may influence the chemical generation reaction itself. The generation $^1O_2$ by photoexcited Rose Bengal has almost no influence of chemical reaction. The Rose Bengal $^1O_2$ generation system was fabricated with super luminescence green LED used for the traffic signal. The $^1O_2$ generation intensity was controlled by monitoring NIR chemiluminescence using the InGaAs-CIA system without liquid nitrogen. **Figure 15** shows a block diagram of the $^1O_2$ generation system.

The Rose Bengal solution was photoexcited by the 5W LED of best match absorption band. The new optical excited system achieved the very small compact size in comparison of using argon ion laser. The system is suitable for chemical plants of mass production because the electric power supply circuits for the LED are simple and easy current control. The generated

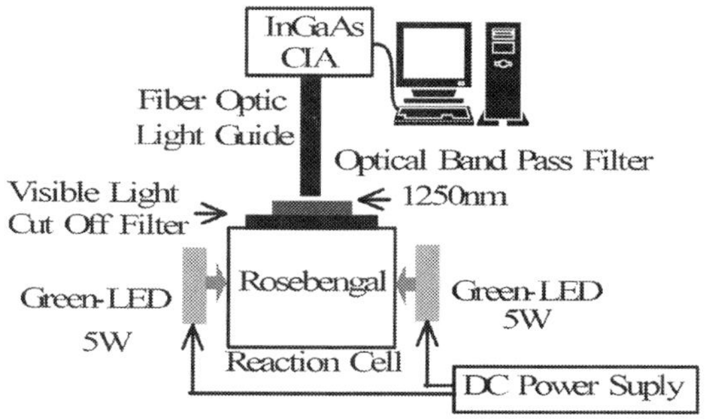

Figure 15. Block diagram of $^1O_2$ generation system by use of Rose Bengal.

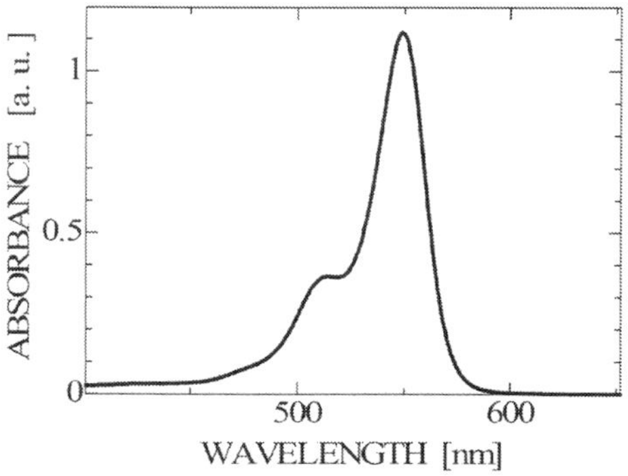

Figure 16. Absorption band of Rose Bengal aqueous.

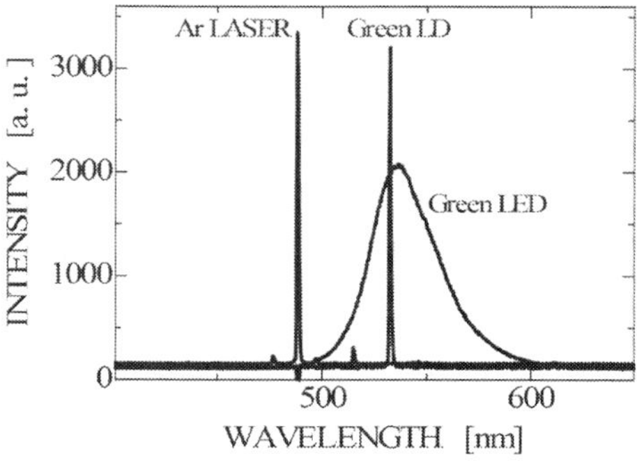

Figure 17. Rose Bengal excitation light source.

http://dx.doi.org/10.5772/66644

$^1O_2$ was optically monitored by InGaAs-CIA at the device temperature of about 150 K degree without liquid nitrogen. The chemical plants should avoid using liquid nitrogen because of choking hazard and troublesome chores. The minimum detectable power of 10^{-13} W at the device temperature 150 K was achieved for monitoring $^1O_2$.

The absorption band of Rose Bengal is shown in **Figure 16**. The absorption band covers from 500–600 nm. The absorption peak spectrum yields at the wavelength of 550 nm. The emission spectrum of an excitation light source is shown in **Figure 17**. A water-cooled multimode argon ion laser or green-laser was used for the traditional optical excitation source. The multimode spectrum of argon ion laser is oscillated at the wavelength of 488 and 515 nm. The intensity of spectrum at 488 nm wavelength is stronger than that of 515 nm. The main spectrum stands at the outside of Rose Bengal absorption band. The spectrum of green LED and super luminosity green LED has the center position of the absorption band. The oscillation power of green super luminosity LED for traffic signal has very strong power in comparison with green LD except for metallic processing green LED. The green super luminosity LED permits increasing of absorbance, and achieved downsizing the system spectacularly, and realized cost cuts. **Figure 18** shows photograph of excitation by green LED.

Figure 18. Photograph of excitation by green LED.

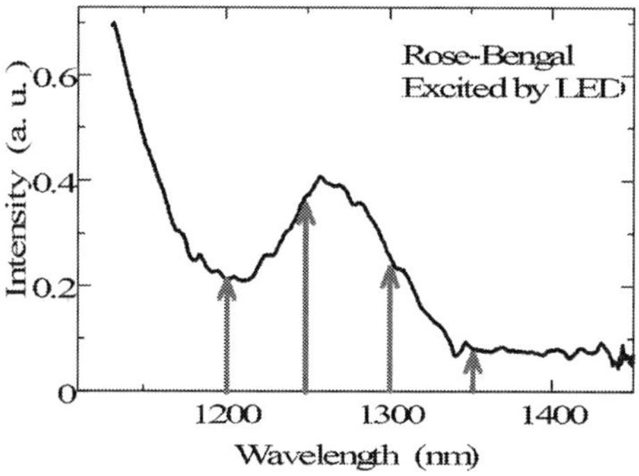

Figure 19. $^1O_2$ spectrum of Rose Bengal excited by green LED.

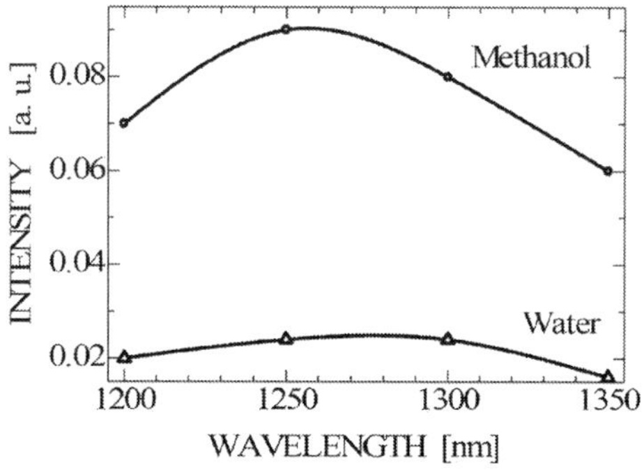

Figure 20. $^1O_2$ spectrum of Rose Bengal by measurement with InGaAs-CIA at 150 K.

Figure 19 shows the spectrum $^1O_2$ of Rose Bengal excited by green LED. The spectrum has a peak at the wavelength of 1.27 μm. A red line in **Figure 19** indicates the wavelength of optical band pass filters. There are four optical band pass filters (1200, 1250, 1300, and 1350 nm) for the spectroscopic system.

The $^1O_2$ emission spectrum of Rose Bengal is shown in **Figure 20**. The intensity of $^1O_2$ emission in methanol is stronger than water solvent. The lifetime of $^1O_2$ in methanol is longer than in water. The generation intensity of $^1O_2$ is monitored through an optical fiber combined with InGaAs-CIA. Many $^1O_2$ monitoring equipment and generation optical sources are necessary in the construction of a microreactor chemical plant. The conventional system employs the Ge-TIA photo detection system with lock-in amplifier and water-cooled argon ion laser. Such a system is not suitable for chemical plant because the system needs occupied large space and high cost. The combination of InGaAs-CIA monitoring and the super luminescence green LED $^1O_2$ generating system enables the construction of a compact and reasonable chemical plant. The system expects to contribute to the development of new medicines.

4. Summary

We developed a highly sensitive InGaAs-CIA in the near-infrared region by use of commercially available photodiode. The system performed low-level light detection measurement in the NIR region. The optical emission of $^1O_2$ was detected specifically in the NIR region. The measurement method of $^1O_2$ was changed from the chemical technique to the physical method. We have achieved detection $^1O_2$ of the heme compound in the living organism including hemoglobin (blood), myoglobin (muscle), and cytochrome c (mitochondria) successfully. These heme compounds play an important role in living body, and also generated $^1O_2$ as a catalyst. The InGaAs-CIA system and multichannel detection system for the low-level light was introduced in the NIR region. The system is available for detection of $^1O_2$ from chemical reaction. The electron refrigeration system without liquid nitrogen will be expected to become popular and realize compact size. Especially, we want to make contribution by means of our system to develop the noninvasive photo-sensitive substance for the photodynamic therapy that produces $^1O_2$ caused oxidative damage to the cancer cells.

Author details

Iwao Mizumoto*, Hiroshi Oguma and Yostumi Yoshi

*Address all correspondence to: mizumoto@nc-toyama.ac.jp

Department of Electronics and Computer Engineering, National Institute of Technology, Toyama College, Toyama, Japan

References

[1] A. U. Kahn, "Near infrared emission of singlet oxygen generated in the dark", J. Biol. Chem., **4**, pp. 200–207 (1989).

[2] D. N. B. Hall, R. S. Joyce and T. W. McCurnin, "Johnson nose limited operation of photovoltaic InSb detectors", Appl. Opt., **14**, pp. 450–453 (1975).

[3] L. E. Dereniac, R. R. Joyce and R. W. Capps, "Low-noise preamplifier for photoconductive detectors", Rev. sci. Instrum., **48**, pp. 392–394 (1977).

[4] F. J. Low, "Integrating amplifiers using cooled J.FETs", Appl. Opt., **23**, pp.1308–1309 (1984).

[5] I. Mizumoto and S. Mashiko, "Low noise near infrared detection system using InGaAs pin photodiode", Elect. Lett., **29**, pp. 234–235 (1193).

[6] I. Miaumoto and S. Mashiko, "Highly sensitive spectroscopic measurement system using an InGaAs pin photodiode in the near-infrared region", Meas. Sci. Technol., **4**, pp. 665–667 (1993).

[7] J. R. Kanofsky, "Singlet oxygen production by bleomycin", J. Biol. Chem., **261**, pp. 13546–13550 (1986).

[8] M. Nakano, T. Noguchi, K. Sugioka, H. Fukuyama, M. Sato, Y. Shimizu, Y. Tsuji and H. Inaba, "Spectroscopic evidence for the generation of singlet oxygen in the reduced nicotinamide adenine dinucleotide phosphate-dependent microsomal lipid peroxidation system", J. Biol. Chem., **250**, pp. 2404–2406 (1975).

[9] N. Suzuki, A. Fujimura, T. Nsgai, I. Mizumoto, T. Itami, H. Hatabe, T. Nozawa, N. Kato, T. Nomoto and B. Yoda, BioFactors, "Antioxidative activity of animal and vegetable dietary fibers", **21**, pp. 329–333 (2004).

Chapter 4

Carbohydrate Analysis by NIRS-Chemometrics

Mercedes G. López,
Ana Sarahí García-González and
Elena Franco-Robles

Additional information is available at the end of the chapter

http://dx.doi.org/10.5772/67208

Abstract

Near-infrared spectroscopy (NIRS) is a high-throughput, low-cost, solvent-free, and nondestructive analytical tool. Chemometrics is the science that employs statistical and mathematical methods to explain near-infrared spectra; it has been proven that when they are coupled, their effectiveness highly improved in-depth carbohydrate characterization. This chapter focuses on the fundamentals of near-infrared spectroscopy in the study of carbohydrates, as well as the application of partial least squares regression (PLSR) and principal component analysis (PCA), as the most useful chemometric techniques involved in carbohydrate analysis. The theoretical aspects and practical applications starting from simple to complex carbohydrates mixtures are covered. Indeed, the contributions from different fields extend the implementation of near-infrared spectroscopy from industrial quality control to scientific research.

Keywords: near-infrared spectroscopy, chemometrics, carbohydrates, polysaccharides, partial least squares regression, principal component analysis

1. Introduction

In a vibrational spectroscopy, near-infrared spectroscopy (NIRS) covers the transition from the visible spectral range to the mid-infrared region. The NIR spectral region ranges from 800 to 2500 nm (12,500–4000 cm^{-1}) with absorptions representing overtones and combinations mainly associated with –CH, –OH, –NH, and –SH functional groups [1]. NIR spectroscopy in combination with chemometric analyses can provide unique information in a wide field of applications from life sciences to environmental issues. It is more frequently used in the agricultural field [2–5], in particular, on the elucidation of nonstructural carbohydrates (NSCs) of plants. NSCs are products of the photosynthesis, providing substrates for growth and metabolism and can be stored by the plant playing a central role in the plant response to the environment [6, 7]. This type of carbohydrates is classified into monosaccharides (glucose and fructose), disaccharides (sucrose), polysaccharides (starch and fructans), oligosaccharides (raffinose), and sugar alcohols (inositol, sorbitol, and mannitol) [8, 9].

NIR spectroscopy is widely used to follow the chemical, physical, technological, or physiological processes that affect the structure and composition of carbohydrates found in many different organisms [10]. The success of this technique relies on the rapid and nondestructive analysis of the sample without the use of chemicals [11]. In addition, the data can be analyzed with chemometric methods. In this regard, partial least squares regression (PLSR) and principal component analysis (PCA) are two of the most recognized statistical methods that can be used to build NIR-chemometric models. PLSR is a well-established method for multivariate modeling and calibration [12]. Meanwhile, PCA analyzes data tables representing observations described by several dependent variables, which are, in general, intercorrelated [13].

The objective of this chapter is to give a comprehensive overview of NIR spectroscopy for analyzing carbohydrates, such as glucose, fructose, sucrose, and fructans. In addition, we describe NIR spectroscopy and multivariate methods used to identify, classify, and quantify carbohydrates in plant tissues. Furthermore, we present the main applications of NIR-chemometrics on carbohydrate analyses.

2. NIR spectra: characteristic bands of oligosaccharide and polysaccharide

The term "near" in NIR relies on the position of the electromagnetic energy lying next to or near the visible energy range. Molecular vibrations in the middle infrared (MIR) range cover absorptions bands between 2500 and 25,000 nm (4000 and 400 cm^{-1}) representing the most intense and simplest bands in the whole infrared range, whereas NIR bands arise in the interval between 800 and 2500 nm (12,500 and 4000 cm^{-1}) covering absorptions corresponding to overtones and combinations of fundamental vibrations [14]. NIR spectroscopy is concerned with both electronic and vibrational transitions [1]. Bands due to electronic transitions are observed in the NIR region and in general are presented as weak bands. Moreover, bands arising from overtones and combination modes are so-called forbidden transitions. Starting from the diatomic molecule as the simplest vibrating system, described by the harmonic and anharmonic oscillator, the study of more complex substances is referred to as polyatomic molecules [14].

The NIR region can be divided into three regions. Region I spans from 800 to 1200 nm (12,500–8500 cm^{-1}), also known as the "the short-wave NIR region (SWNIR)," "near-NIR region (NNIR)," or "the Herschel region," represents bands resulting from electronic transitions, overtones, and combinations modes. Region II ranges from 1200 to 1800 nm (8500–5500 cm^{-1}) and covers first overtones of XH (X = C, O, N), stretching vibrations and various types of combination modes. Finally, Region III (1800–2500 nm or 5500–4000 cm^{-1}) is a combination mode region. Many applications utilize Regions II and III [1].

Absorptions due to different functional groups, especially –CH, –OH, and –NH, are displayed as molecular overtones and combination vibrations at specific wavebands [15, 16]. NIR spectral data are influenced by a particle size (e.g., ground or powder) and need to be properly calibrated [17]. In **Table 1**, the characteristic bands of oligosaccharide and polysaccharide are listed.

NIRS has been used as a fingerprint technique for all kinds of samples (liquids, solids, and semisolids), independently of their nature, relatively simple substances or pure compounds, most times they show broad and overlapping bands, it is impossible to correctly assign the specifically vibrations, and cannot be used for structural determination of a sample [18].

Carbohydrate type	Waveband	Wavenumber		Reference
		nm	cm^{-1}	
Glucose	OH stretch 1st overtone	2340, 2255, 2150, 2085, 1902, 1730, 1590, 1520, 1385, 1195	4274, 4435, 4651, 4796, 5258, 5780, 6289, 6579, 7220, 8368	[19]
Glucose	OH stretch/OH bend	1688	5924	[20]
Sucrose	OH stretch 1st overtone	1433	6978	[20]
Sucrose/glucose/fructose	OH combination	1928	5186	[21]
Sucrose/glucose/fructose	OH stretch/CO stretch combination	2123–2200	4710–4717	[22]
Crystalline sucrose	OH stretch 1st overtone	1443–1440	6930–6944	[23, 24]
Polysaccharides	CH stretch/CH deformation combination	2328	4295	[25]
Polysaccharides	OH stretch/CO stretch combination	2274, 2271–2270	4398, 4403–4405	[22, 25]
Polysaccharides	OH combination	2090	4785	[25]
Polysaccharides	OH stretch/OH bend	1920	5208	[25]
Polysaccharides	CH combination/CH 1st overtone	2328, 2270, 2078, 1920, 1587–1583	4295, 4405, 4813, 5208, 6300–6317	[19, 22]
Polysaccharides	OH stretch 1st overtone	1437–1389	6960–7200	[26–29]

Table 1. Characteristic bands of oligosaccharide and polysaccharide.

3. Multivariate data analysis by NIRS

NIR spectra are characterized for their complexity and difficulty to be interpreted. For these reasons, multivariate methods from chemometrics are required to understand NIR spectra.

Chemometrics comprise the development and use of mathematical and statistical methods for applications in chemistry. As a discipline, the aim of chemometrics is to provide methods to extract relevant chemical information out of measured chemical data in order to represent and display this information.

Figure 1 shows a general scheme for multivariate techniques, including the two different chemometric groups that are frequently employed in the NIR spectra analysis: the qualitative (classification) methods and the quantitative (regression) methods. As a first step, before choosing any method, usually NIR spectra are preprocessed with mathematical treatments, such as baseline correction, normalizations, derivatives, and smoothing, in order to enhance the relevant information and reduce the influence of side information contained in the spectra. The classification methods are used to group or separate the samples according to their spectra. The regression methods correlate the spectrum to quantifiable properties of the samples.

3.1. Quantitative analysis

The basic principles used for quantitative analysis are fundamentally invariable for all optical and spectral measurement methods. The principle behind any quantitative analysis is that the desired quantity, property, parameter, or compound can be determined from the signal obtained by an instrument, and this signal differs in a predictable manner for a given experimental system. The magnitude of the signal obtained can be correlated, directly or by mathematical algorithms, to the target characteristic properties of a sample. A common implementation of quantitative analysis is the determination of the concentration of a given analyte. For most applications, an attempt is made to linearize the relationship between the analyte and the instrument response, although this is not essential if a well-defined mathematical relationship can be established. This leads to the generation of a calibration from a characterized standard set (references) with the objective to construct a prediction model for a group of samples (**Figure 2**) [30].

Many successful NIRS analysis have been performed using PLSR as a quantitative chemometric technique. Its usefulness derives from its potential to analyze data with numerous, noisy, collinear, and even incomplete variables. By establishing a linear relationship between two

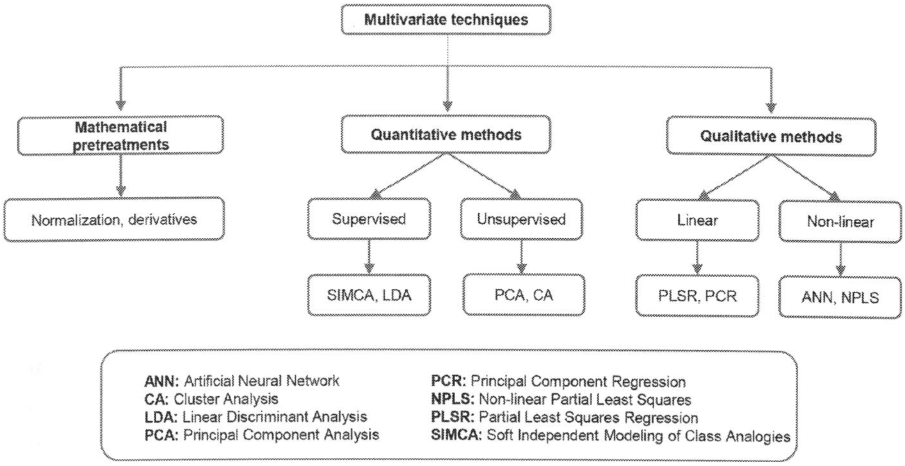

Figure 1. General scheme showing the commonly multivariate techniques employed by NIR spectroscopy.

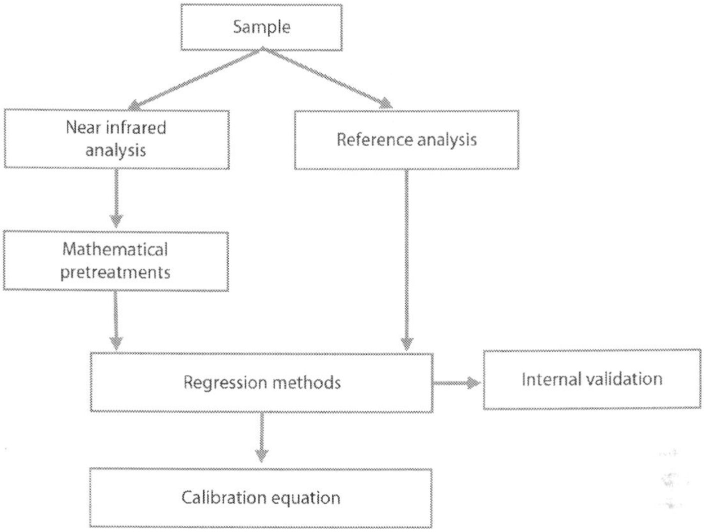

Figure 2. Scheme for the construction of a quantitative model.

data matrices, the spectral data X and the reference values Y, through a linear multivariate model, the PLSR technique finds out the variables in the X matrix that will best define the Y matrix. In other words, it represents the NIR spectra in the space of wavelengths in order to display directions that will be linear combinations of wavelengths called factors that describe the studied property [31, 32].

3.2. Qualitative analysis

Qualitative analyses are used for the classification of samples in accordance with their NIR spectra. Two general approaches can be used for qualitative classification: the unsupervised and the supervised methods. In the first approach, samples are classified lacking preceding knowledge, except the spectra. On the other hand, supervised methods require a prior knowledge of the sample, for instance, a category membership, generating a classification model with a training set of samples with well-established categories. The obtained model performance is evaluated by relating the classification predictions to the well-known categories of the validation samples [33].

Principal component analysis (PCA) is one of the most popular classification methods utilized in life sciences. PCA is used to visualize the most important information from a given data. One of the most significant advantages of PCA application is the reduction of the number of variables (scores), allowing the representation of a multivariate data table in a small dimensional area. Its purpose is to obtain significant information from the NIR spectra to express it as a set of new orthogonal variables called principal components (PC). The first principal component (PC1) defines the maximum variability scattered within the samples. A second principal component (PC2), uncorrelated and orthogonal to the first principal component, explains the maximum variability not described by the first component, this behavior continues with the next principal component (PC3), and so on [12]. Thereby, a display pattern of similarity of the variables as points in maps is created.

4. Applications of carbohydrates analysis by NIRS

The near-infrared spectroscopy (NIRS) is a technique that allows the measurement of carbohydrates in a wide variety of samples. Nowadays, NIRS-chemometrics have proven their effectiveness for both qualitative and quantitative carbohydrate analysis. NIRS has several advantages such as allowing the sample remains intact after analysis and giving access to multiple chemical as well as physical properties at the same time [34].

NIR spectroscopy is generally chosen for its high-throughput screening, reduced sample preparation, low cost, and the nondestructive nature toward the analyzed sample [14]. However, establishing a suitable calibration demands a big effort and requires reference values for each sample, which makes it time-consuming and costly at the beginning [35].

In the agrifood sector, the potential of NIRS have been widely investigated, this is a very powerful tool that provides meaningful information about internal and external properties of fruits, such as sugar content, total acidity, pH, soluble solid content, dry matter, firmness, and bruises, to mention some [36]. Moreover, NIRS can be applied to a wide variety of problems such as determination of particle size [38], determination of the best harvesting time [37], and investigation of geographical origin of foods such as apples, meat, and cheese [39].

However, and particularly to specific sugar content, NIRS in combination with PLSR models has been used in sorghum stalks [40] and sweet sorghum (cellulose, lignin, and hemicellulose) [41], fruit juices [42, 43], rice (amylose) [44], whey (lactose) [45], grasses (fructans) [46, 47], maize (nonstructural and water soluble carbohydrates) [48], intact apple fruit to determinate fructose, glucose, and sucrose [49], orange [50], apricot [51], sugar beet [52], cherries [53], and other fruits (**Table 2**). All these studies accorded that the performance of NIR spectroscopy is comparable to the reference chromatographic method, but the former is much faster and easier to carry out.

On the other hand, NIRS has been applied on food quality evaluation; it is often used to check if fruits or vegetables are green or rotten to detect surface defects. NIRS is also employed to check sugar concentrations, for instance, not only in apples [64], oranges [55, 56], mango [65], kiwifruits [57], sugar beet [54], peaches [66], jujube [67], onion [68], potato tubers [58], Nules Clementine [62], passion fruit [69], but also in fruit juices [43], wine [59], or cakes [60] (**Table 2**). Additionally, it has been used in breadstuff, dairy products, meat, vegetables, and fish products and in processed food to provide information about overtones and their combinations [70]. Moreover, studies have been performed to demonstrate that NIRS-chemometric analyses are of greater predictive value than mid-infrared data. In Chinese yams, Zhuang et al. [63] analyzed with NIR and MIR spectroscopy, the authors concluded that reasonable results were obtained using both spectral data sets and methods, but that NIR-chemometric data derived better prediction models.

Sample	Carbohydrate	Analysis	Reference
Grain sorghum stalks	Sucrose, glucose	PLSR	[41]
Fruit juices	Glucose, fructose, sucrose	PLSR, PCA	[43, 44]
Rice	Amylose	mPLSs	[45]
Whey	Lactose	PLS	[46]
Grasses	Fructan	PLSR	[47, 48]
Apple fruit	Glucose, fructose, sucrose	PLS	[50]
Sugar beet	Sucrose	SEPs	[53, 55]
Cherries	Total carbohydrates	PLSR	[54]
Oranges	Glucose, fructose, sucrose	PLSR	[56, 57]
Kiwifruit	Glucose, fructose, sucrose	PLS	[58]
Potato	Glucose, fructose	PLSR	[59]
Wine	Glucose	PLSR, PCR	[60]
Cakes	Sucrose	MLR	[61]
Syrup	Glucose, fructose, sucrose	PLSR	[62]
Nules Clementine	Glucose, fructose, sucrose total carbohydrates	PLS, PCR	[63]
Chinese yams	Total carbohydrates	PLS, PCA, LS-SVM	[64]

mPLSs, various modified partial least square; PLS, partial least square; SEPs, standard errors of prediction; MLR, multiple linear regressions; and LS-SVM, least squares-support vector machine.

Table 2. Samples analyzed by NIR in a carbohydrate study.

In respect to specific absorption peaks, sugar analyses have been carried out in fruit juices establishing that NIRS can deal with the distortions due to water clusters [20–22, 42].

NIR techniques have also been applied to measure biomass composition, especially on the presence of structural carbohydrates. The National Renewable Energy Laboratory (NREL) reported sorghum composition prediction models for glycan, xylan, lignin, starch, extractives, and ash [71].

NIR spectroscopy is not only useful in laboratory measurements sites but also applicable to online and field studies. The study of 116 syrup samples to compare a portable spectrometer and a benchtop device showed that the reduced wavelength range and reduced resolution of the portable device is sufficient to receive calibrations with $R^2 \geq 0.96$ for standard syrups with comparable standard error of prediction (SEP) values of 1.30 g/100 g versus 1.19 g/100 g, 0.94 g/100 g versus 0.99 g/100 g, and 2.04 g/100 g versus 2.46 g/100 g for glucose, fructose, and sucrose, respectively, to the handheld device [61]. The developed method is suitable to be implemented for quality control in the producing industry as well as in grocery stores.

A relevant novel application of the predictive models, particularly of the direct NIR prediction on diverse parameters on fruit quality was demonstrated. In Ref. [50], the authors compared two commercial portable spectrometers (Vis/NIR spectrometer versus OTF-NIR) for four orange varieties quality: soluble solids content, acidity, titratable acidity, maturity index, flesh firmness, juice volume, fruit weight, rind weight, juice volume to fruit weight ratio, fruit color index, and juice color index, and they found relevant the prediction of maturity index. The Lab spec spectrometer showed better predictive performance than the laminar instrument.

In another study, a Lab spec Pro portable spectrophotometer to conduct an online classification of beef tenderness was also successful [72].

In sugar-flour mixtures, NIR spectroscopy displayed proper results on the characteristic absorption bands of sugars, which are 1200 nm (8333 cm^{-1}), 1437 nm (6959 cm^{-1}), 2074 nm (4822 cm^{-1}), and 2320 nm (4310 cm^{-1}). However, it was not possible to distinguish various sorts of sugars, for instance, make a difference between the sucrose of the powdered sugar and the numerous carbohydrates present in the flour. Nevertheless, the identification of specific signatures of sugars can be very useful for rapid detection in the industrial sector [73].

Honey represents another class of samples that have proven the effectiveness of a NIR analysis [74]. In a study on Galicia honeys with protected geographical indication (PGI), the samples were processed by different chemometric methods to develop an authentication system specific to this type of honey. In this work, fifteen Galicia certificated PGI honeys were differentiated from other fifteen commercial available honeys by PCA, demonstrating that a single and fast chemometric method could be used to indicate the genuineness of Galicia PGI samples. **Figure 3A** shows the NIR spectra of all the analyzed samples and **Figure 3B**, illustrates the discrimination of Galicia PGI honeys from the other samples by the PCA plot.

Similarity, the potential use of NIR-PCA analysis to monitor sugar adulteration in onion powders was assessed through a detailed examination of the feasibility of quantification of cornstarch as an adulterating ingredient in onion powders [75]. Spectral analysis of 18 concentrations of starch in 180 onion powders, ranging from 0 to 35%, was conducted. The NIR spectra of the pure and adulterated onion powders (**Figure 4A**) reveal differences in

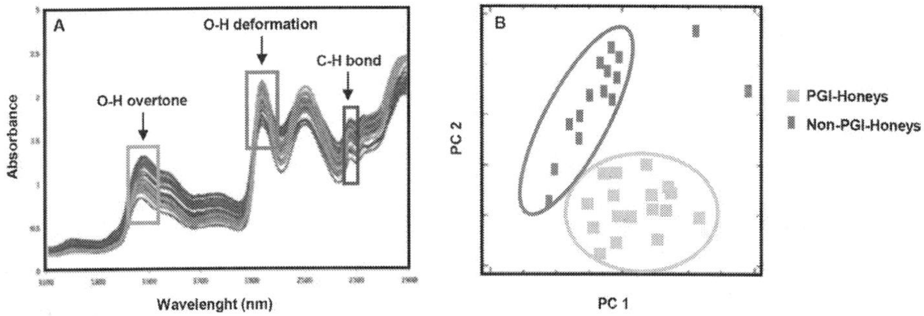

Figure 3. (A) NIR spectra of honey samples. (B) Score plot of the honey samples in the space defined by the first two principal components. Adapted from Ref. [74].

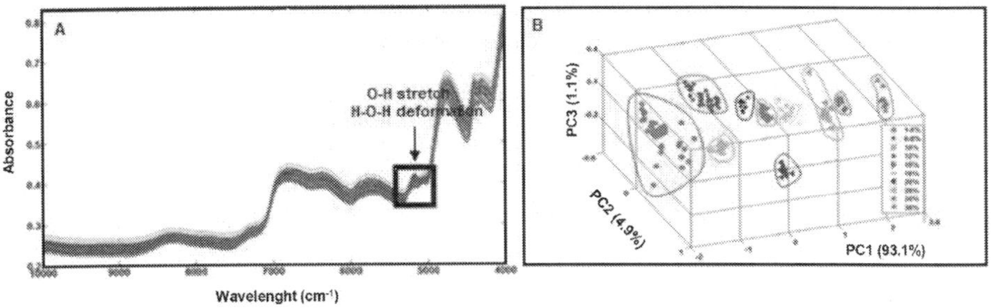

Figure 4. (A) Original NIR spectra of pure onion and starch onion mixtures at different concentrations. (B) Principal component score plot for the first three PCs for discrimination among different adulteration concentrations in onion powder. Adapted from Ref. [75].

the absorption intensities between 1920 (5208 cm^{-1}) and 1980 nm (5051 cm^{-1}). The absorption bands in these regions correspond to the O–H stretch and O–H band combination and the H–O–H deformation combination, which represents the starch content. The spectral fluctuations from 1400 (7143 cm^{-1}) to 1600 nm (6250 cm^{-1}) correspond to the first overtone of the hydroxyl group. The precise position of these bands is very sensitive to hydrogen bonding in the starch molecule, causing a difference between genuine and adulterated samples. The application of PCA (**Figure 4B**) resulted in a data grouping of each of the different concentrations used, working as a discriminative screening tool of authentic and adulated samples.

Applications of NIRS have been developed also in the nutrition and health fields. NIR and MIR spectroscopy measurements and multivariate calibration methods based on partial least square regression have been used in a determination of fat, proteins, carbohydrates, and energy values in baby food, infant fast food, and canteen menus, with a simple, fast, and good predictive capabilities [70]. Another great diagnostic application is the measurement of blood glucose [1].

Finally, another notable capacity of NIRS was the prediction of carbohydrates concentrations, and distribution, leading to high ratio of performance to deviation (RPD) values, reducing the use of chemicals and working time, confirming that this makes a suitable technique of industry applications [61].

5. Conclusions

The potential of NIR spectroscopy in combination with chemometrics on carbohydrate analysis has been fully demonstrated. NIR is a powerful technique to study carbohydrates composition, type, and levels. This method can be used qualitatively and quantitative to detect, identify, and qualify carbohydrates. These unique capabilities enable the employment of NIR-chemometric in numerous applications: from state-of-the-art scientific experiments to on-line industrial processing control.

Author details

Mercedes G. López[1]*, Ana Sarahí García-González[1] and Elena Franco-Robles[2]

*Address all correspondence to: mlopez@ira.cinvestav.mx

1 Center for Research and Advanced Studies of the National Polytechnic Institute, Irapuato, Mexico

2 University of Guanajuato, Irapuato, Mexico

References

[1] Ozaki Y. Near-infrared spectroscopy—its versatility in analytical chemistry. Analytical Sciences: the International Journal of the Japan Society for Analytical Chemistry. 2012;**28**:545–563. DOI: 10.2116/analsci.28.545.

[2] Kelley SS, Rials TG, Snell R, Groom LH, Sluiter A. Use of near infrared spectroscopy to measure the chemical and mechanical properties of solid wood. Wood Science and Technology. 2004;**38**:257–276. DOI: 10.1007/s00226-003-0213-5.

[3] Ono K, Hiraide M, Amari M. Determination of lignin, holocellulose, and organic solvent extractives in fresh leaf, litter fall, and organic material on forest floor using near-infrared reflectance spectroscopy. Journal of Forest Research. 2003;**8**:191–198. DOI: 10.1007/s10310-003-0026-2.

[4] Philip YX, Liu L, Hayes D, Womac A, Hong K, Sokhansanj S. Fast classification and compositional analysis of corns over fractions using Fourier transform near-infrared techniques. Bioresource Technology. 2008;**99**:7323–7332. DOI: 10.1016/j.biortech.2007.12.063.

[5] Jin S, Chen H. Near-infrared analysis of the chemical composition of rice straw. Industrial Crops and Products. 2007;**26**:207–211. DOI: 10.1016/j.indcrop.2007.03.004.

[6] Chapin I, Schulze A, Mooney HA. The ecology and economics of storage in plants. Annual Review of Ecology and Systematics. 1990;**21**:423–447. DOI: 10.1146/annurev.es.21.110190.002231.

[7] Kozlowski TT. Carbohydrate sources and sinks in woody plants. The Botanical Review. 1992;**58**:107–222. DOI: 10.1007/BF02858600.

[8] Rastall RA. Methods in Plant Biochemistry. Vol. 2. Academic Press. London. 1990. DOI: 10.1016/0968-0004(91)90104-4.

[9] Stick R, Williams S. Carbohydrates: The Essential Molecules of Life. 2nd ed. Oxford: Elsevier. 2009. DOI: 10.1016/j.carbpol.2010.04.058.

[10] Cassells J, Reuss R, Osborne B, Wesley I. Near infrared spectroscopic studies of changes in stored grain. Journal of Near Infrared Spectroscopy. 2007;**15**:161–167. DOI: 10.1255/jnirs.727.

[11] Jiang W, Han G, Zhang Y, Wang M. Fast compositional analysis of ramie using near-infrared spectroscopy. Carbohydrate Polymers. 2010;**81**:937–941. DOI: 10.1016/j.carbpol.2010.04.009.

[12] Abdi H, Williams LJ. Principal component analysis. Wiley Interdisciplinary Reviews: Computational Statistics. 2010;**2**:433–459. DOI: 10.1002/wics.101.

[13] Geladi P, Kowalski BR. Partial least-squares regression: a tutorial. Analytica Chimica Acta. 1986;**185**:1–17. DOI: 10.1002/wics.101.

[14] Siesler HW, Ozaki Y, Kawata S, Heise HM. Near-Infrared Spectroscopy: Principles, Instruments, and Applications. Wiley: Weinheim. 2008:361 p. DOI: 10.1002/9783527612666.

[15] Reich G Near-infrared spectroscopy and imaging: basic principles and pharmaceutical applications. Advanced Drug Delivery Reviews. 2005;**57**:1109–1143. DOI: 10.1016/j.addr.2005.01.020.

[16] Osborne BG. Near-infrared spectroscopy in food analysis. Encyclopedia of Analytical Chemistry. Meyers RA: Wiley, Chichester. 2006. pp. 1–14. DOI: 10.1002/9780470027318.a1018.

[17] Pasikatan MC, Steele JL, Spillman CK, Haque E. Near infrared reflectance spectroscopy for online particle size analysis of powders and ground materials. Journal of Near Infrared Spectroscopy. 2001;**9**:153–164. DOI: 10.1255/jnirs.303.

[18] Huck CH. Advances of vibrational spectroscopic methods in phytomics and bioanalysis. Journal of Pharmaceutical and Biomedical Analysis. 2014;**87**:26–35. DOI: 10.1016/j.jpba.2013.05.010.

[19] Ghosh S, Roy RB. Quantitative near-infra-red analysis of reducing sugar from the surface of cotton. The Journal of Textile Institute. 1998;**79**:504–510. DOI: 10.1155/2013/649407.

[20] Giangiacomo R, Magee JB, Birth GS, Dull GG. Predicting concentrations of individual sugars in dry mixtures by near-infrared reflectance spectroscopy. Journal of Food and Science. 1981;**46**:531–534. DOI: 10.1111/j.1365-2621.1981.tb04903.x.

[21] Giangiacomo R.. Study of water–sugar interactions at increasing sugar concentration by NIR spectroscopy. Food Chemistry. 2006;**96**:371–379. DOI: 10.1016/j.foodchem.2005.02.051.

[22] Rambla FJ, Garrigues S, de la Guardia M. PLS-NIR determination of total sugar, glucose, fructose and sucrose in aqueous solutions of fruit juices. Analytica Chimica Acta. 1997;**344**:41–53. DOI: 10.1016/S0003-2670(97)00032-9.

[23] Davies AMC, Miller CE. Tentative assignment of the 1440-nm absorption band in the near-infrared spectrum of crystalline sucrose. Applied Spectroscopy. 1998;**42**:703–704. DOI: 10.1366/0003702884429364.

[24] Workman Jr J, Weyer L. Practical Guide to Interpretive Near-Infrared Spectroscopy. New York, NY: CRC Press. 2007:344 p. DOI: 10.1002/anie.200885575.

[25] Shenk JS, Workman JJ, Westerhaus MO. Application of NIR spectroscopy to agricultural products. In: Handbook of Near-Infrared Analysis. 2nd ed. CRC Press. New York. Basel. 2001. pp. 419–470. DOI: 10.1201/9781420002577.pt4a.

[26] Tsuchikawa S, Siesler HW. Near-infrared spectroscopic monitoring of the diffusion process of deuterium-labeled molecules in wood. Part I: softwood. Applied Spectroscopy. 2003;**57**:667–674.

[27] Tsuchikawa S, Siesler HW. Near-infrared spectroscopic monitoring of the diffusion process of deuterium-labeled molecules in wood. Part II: hardwood. Applied Spectroscopy. 2003;**57**:675–681.

[28] Krongtaew C, Messner K, Ters T, Fackler K. Characterization of key parameters for biotechnological lignocellulose conversion assessed by FT-NIR spectroscopy. Part I. Qualitative analysis of pretreated straw. BioResources. 2010;**5**:2063–2080.

[29] Krongtaew C, Messner K, Ters T, Fackler K. Characterization of key parameters for biotechnological lignocellulose conversion assessed by FT-NIR spectroscopy. Part II. Quantitative analysis by partial least squares regression. BioResources. 2010;**5**:2081–2096.

[30] Coates J. Classical methods of quantitative analysis. In: Chalmers JM, Griffiths PR, editors. Handbook of Vibrational Spectroscopy. 1st ed. Wiley. New Jersey. 2006. pp. 2235–2237. DOI:10.1002/0470027320.s4602.

[31] Wold S, Sjöström M, Eriksson L. PLS-regression: a basic tool of chemometrics. Chemometrics and Intelligent Laboratory Systems. 2001;**58**:109–130. DOI: 10.1016/S0169-7439(01)00155-1.

[32] Roggo Y, Chalus P, Maurer L, Lema-Martinez C, Edmond A, Jent N. A review of near infrared spectroscopy and chemometrics in pharmaceutical technologies. Journal of Pharmaceutical and Biomedical Analysis. 2007;**44**:683–700. DOI: 10.1016/j.jpba.2007.03.023.

[33] Lavine BK. Chemometrics. Analytical Chemistry. 2000;**72**:91R–97R. DOI: 10.1021/a1000016x.

[34] Tsuchikawa S. A review of recent near infrared research for wood and paper. Applied Spectroscopy Reviews. 2007;**42**:43–71. DOI: 10.1080/05704920601036707

[35] Bakeev KA. Process Analytical Technology: Spectroscopic Tools and Implementation Strategies for the Chemical and Pharmaceutical Industries. 2nd ed. John Wiley & Sons, Ltd. New Jersey. 2010. 543 p. DOI: 10.1002/9780470689592.

[36] Lin H and Ying Y. Theory and application of near infrared spectroscopy in assessment of fruit quality: a review. Sensing and Instrumentation for Food Quality and Safety. 2009;**3**:130–141. DOI: 10.1007/s11694-009-9079-z.

[37] Bittner LK, Heigl N, Petter CH, Noisternig MF, Griesser UJ, Bonn GK, Huck CW. Near-infrared reflection spectroscopy (NIRS) as a successful tool for simultaneous identification and particle size determination of amoxicillin trihydrate. Journal of Pharmaceutical and Biomedical Analyses. 2011;**54**:1059–1064. DOI: 10.1016/j.jpba.2010.12.019.

[38] Huck CW. Advances of infrared spectroscopy in natural product research. Phytochemistry Letters. 2015;**11**:384–393. DOI: 10.1016/j.phytol.2014.10.026.

[39] Huck-Pezzei V. Alps food authentication, typicality and intrinsic quality by near infrared spectroscopy. Food Research International. 2014;**62**:984–990. DOI: 10.1016/j.foodres.2014.05.021.

[40] Chen SF, Danao MGC, Singh V, Brown PJ. Determining sucrose and glucose levels in dual-purpose sorghum stalks by Fourier transform near infrared (FT-NIR) spectroscopy. Journal of the Science and Food Agriculture. 2014;**94**:2569–2576. DOI: 10.1002/jsfa.6606.

[41] Wu L, Li M, Huang J, Zou W, Hu S, Li Y, Fan C, Zhang R, Jing H, Peng L, Feng S. A near infrared spectroscopic assay for stalks soluble sugars, bagasse enzymatic saccharification and wall polymers in sweet sorghum. Bioresource Technology. 2015;**177**:118–124. DOI: 10.1016/j.biortech.2014.11.073.

[42] Rodriguez-Saona LE, Fry FS, McLaughlin MA, Calvey EM. Rapid analysis of sugars in fruit juices by FT-NIR spectroscopy. Carbohydrate Research. 2001;**336**:63–74. DOI: 10.1016/S0008-6215(01)00244-0.

[43] Jha SN, Gunasekaran S. Authentication of sweetness of mango juice using Fourier transforms infrared-attenuated total reflection spectroscopy. Journal of Food Engineering. 2010;**101**:337–342. DOI: 10.1016/j.jfoodeng.2010.07.019.

[44] Bagchi TB, Sharma S, Chattopadhyay K. Development of NIRS models to predict protein and amylose content of brown rice and proximate compositions of rice bran. Food Chemistry. 2016;**191**:21–27. DOI: 10.1016/j.foodchem.2015.05.038.

[45] Kucheryavskiy S, Lomborg CJ. Monitoring of whey quality with NIR spectroscopy—a feasibility study. Food Chemistry. 2015;**176**:271–277. DOI: 10.1016/j.foodchem.2014.12.086.

[46] Shetty N, Gislum R. Quantification of fructan concentration in grasses using NIR spectroscopy and PLSR. Field Crops Research. 2011;**120**:31–37. DOI: 10.1016/j.fcr.2010.08.008.

[47] Shetty N, Gislum R, Jensenb AMB, Boelta B. Development of NIR calibration models to assess year-to-year variation in total non-structural carbohydrates in grasses using PLSR. Chemometrics and Intelligent Laboratory Systems. 2012;**111**:34–38. DOI: 10.1016/j.chemolab.2011.11.004.

[48] Campo L, Monteagudo AB, Salleres B, Castro P, Moreno-Gonzalez J. NIRS determination of non-structural carbohydrates, water-soluble carbohydrates and other nutritive quality traits in whole plant maize with wide range variability. Spanish Journal of Agriculture Research. 2013;**11**:463–471. DOI: 10.5424/sjar/2013112-3316.

[49] Liu Y, Ying Y, Yu H, and Fu X. Comparison of the HPLC method and FT-NIR analysis for quantification of glucose, fructose, and sucrose in intact apple fruits. Journal of Agricultural and Food Chemistry. 2006;**54**:2810–2815. DOI: 10.1021/jf052889e.

[50] Cayuela JA, Weiland C. Intact orange quality prediction with two portable NIR spectrometers. Postharvest Biology and Technology. 2010;**58**:113–120. DOI: 10.1016/j.postharvbio.2010.06.001.

[51] Camps C, Christen D. Non-destructive assessment of apricot fruit quality by portable visible-near infrared spectroscopy. LWT - Food Science and Technology. 2009;**42**:1125–1131. DOI: 10.1016/j.lwt.2009.01.015.

[52] Pan L, Zhu Q, Lu R, McGrath JM. Determination of sucrose content in sugar beet by portable visible and near-infrared spectroscopy. Food Chemistry. 2015;**167**:264–271. DOI: 10.1016/j.foodchem.2014.06.117.

[53] Lu R Predicting firmness and sugar content of sweet cherries using near–infrared diffuse reflectance spectroscopy. Transactions of the ASAE. 2001;**44**:1265–1271. DOI: 10.13031/2013.6421.

[54] Roggo Y, Duponchel L, Huvenne J. Quality evaluation of sugar beet (beta vulgaris) by near-infrared spectroscopy. Journal of Agricultural and Food Chemistry. 2004;**52**:1055–1061. DOI: 10.1021/jf0347214.

[55] Li W, Goovaerts P, Meurens M. Quantitative analysis of individual sugars and acids in orange juices by NIR spec. of dry extract. Journal of Agricultural and Food Chemistry. 1996;**44**:2252–2259. DOI: 10.1021/jf9500750.

[56] Luo C, Long X, Liu M, Li J, Wang X. Nondestructive measurement of sugar content in navel orange based on Vis-NIR spectroscopy. In: IFIP international federation for information processing, 4th IFIP TC 12 conference, CCTA; Computer and Computing Technologies in Agriculture IV, vol. Part 1, no. Springer; 2011. pp. 467–473. DOI: 10.1007/978-3-642-18369-0_55.

[57] Slaughter DC, Crisosto CH. Nondestructive internal quality assessment of kiwifruit using near-infrared spectroscopy. Seminars in Food Analysis. 1998;**3**:131–140.

[58] Chen JY, Zhang H, Miao Y, Asakura M. Nondestructive determination of sugar content in potato tubers using visible and near infrared spectroscopy. Japan Journal of Food Engineering. 2010;**11**:56–64. DOI: 10.11301/jsfe.11.59.

[59] Martelo-Vidal MJ, Vazquez M. Evaluation of ultraviolet, visible, and near infrared spectroscopy for the analysis of wine compounds. Czech Journal of Food Sciences. 2014;**32**:37–47.

[60] Osborne BG, Fearn T, Randall PG. Measurement of fat and sucrose in dry cake mixes by near infrared reflectance spectroscopy. Journal of Food Technology. 1983;**18**:651–656. DOI: 10.1111/j.1365-2621.1983.tb00304.x.

[61] Hen R, Schwab A, Huck CW. Evaluation of benchtop versus portable near-infrared spectroscopic method combined with multivariate approaches for the fast and simultaneous quantitative analysis of main sugars in syrup formulations. Food Control. 2016;**68**:97–104. DOI: 10.1016/j.foodcont.2016.03.037.

[62] Magwasa LS, Landahl S, Cronje PJR, Nieuwoudt HH, Mouazen AM, Nicolai BM, Terry LA, Opara UL. The use of Vis/NIRS and chemometric analysis to predict fruit defects and postharvest behavior of "Nules Clementine" mandarin fruit. Food Chemistry. 2014;**163**:267–274. DOI: 10.1016/j.foodchem.2014.04.085.

[63] Zhuang H, Ni Y, Kokot S. A comparison of near- and mid-infrared spectroscopic methods for the analysis of several nutritionally important chemical substances in the Chinese Yam (*Dioscorea opposita*): total sugar, polysaccharides, and flavonoids. Applied Spectroscopy. 2015;**69**:488–495. DOI: 10.1366/14-07655.

[64] Bobelyn E, Serban AS, Nicu M, Lammertyn J, Nicola¨ı BM, Saeys W. Postharvest quality of apple predicted by NIR-spectroscopy: study of the effect of biological variability on spectra and model performance. Postharvest Biology and Technology. 2010;**55**(3):133–143. DOI: 10.1016/j.postharvbio.2009.09.006.

[65] Munawar AA, Budiastra IW. Non-destructive inner quality prediction in intact mango with NIR spectroscopy. In: Beyerer J, Puente León F, Laengle T, editors. Optical Characterization of Materials—conference proceedings. 2015.

[66] Carlomagno G, Capozzo L, Attolico G, Distante A. Non-destructive grading of peaches by near-infrared spectrometry. Infrared Physics & Technology. 2004;**46**:23–29. DOI: 10.1016/j.infrared.2004.03.004.

[67] Zhang S Robust model of fresh jujube soluble solids content with near infrared (NIR) spectroscopy. African Journal of Biotechnology. 2012;**11**:8133–8140. DOI: 10.5897/AJB12.049.

[68] Vincke D, Baeten V, Sinnaeve G, Dardenne P, Fernández-Pierna JA. Determination of outer skin in dry onions by hyperspectral imaging spectroscopy and chemometrics. NIR News. 2014;**25**:9–12. DOI: 10.1255/nirn.1425.

[69] Maniwara P, Nakano K, Boonyakiat D, Ohashi S, Hiroi M, Tohyama T. The use of visible and near infrared spectroscopy for evaluating passion fruit postharvest quality. Journal of Food Engineering. 2014;**143**:33–43. DOI: 10.1016/j.jfoodeng.2014.06.028.

[70] Cascant MM, Garrigues S, de la Guardia M. Direct determination of major components in human diets and baby foods. Analytical and Bioanalytical Chemistry. 2015;**407**:1961–1972. DOI: 10.1007/s00216-015-8461-4.

[71] Wolfrum E, Payne C, Stefaniak T, Rooney W, Dighe N, Bean B, Dahlberg J. Multivariate calibration models for sorghum composition using near-infrared spectroscopy. Technical Report NREL/TP-5100-56838. 2013:1–7. DOI: 10.1007/s00216-015-8461-4.

[72] Shackelford SD, Wheeler TL, Koohmaraie M. On-line classification of US Select beef carcasses for longissimus tenderness using visible and near-infrared reflectance spectroscopy. Meat Science. 2005;**69**:409–415. DOI: 10.1016/j.meatsci.2004.08.011.

[73] Boulley L, Henning S, Pierson JF. Near-infrared optical spectroscopy of sugar-based mixtures—a snapshot to identify issues of influence. In: OCM 2015, 2nd International Conference on Optical Characterization of Materials; March 18th–19th, 2015; Karlsruhe, Germany. 2015. pp. 27–37. DOI: 10.5445/KSP/1000044906.

[74] Herrero-Latorre C, Pena-Crecente RM, Garcia-Martin S, Barciela- Garcia J. A fast chemometric procedure based on NIR data for authentication of honey with protected geographical indication. Food Chemistry. 2013;**141**:3559–3565. DOI: 10.1016/j.foodchem.2013.06.022.

[75] Lohumi S, Lee S, Lee WH, Kim MS, Mo C, Bae H, Cho BK. Detection of starch adulteration in onion powder by FT-NIR and FT-IR spectroscopy. Journal of Agricultural and Food Chemistry. 2014;**62**:9246–9251. DOI: 10.1021/jf500574m.

Chapter 5

Using Near-Infrared Spectroscopy in Agricultural Systems

Francisco García-Sánchez, Luis Galvez-Sola,

Juan J. Martínez-Nicolás,

Raquel Muelas-Domingo and Manuel Nieves

Additional information is available at the end of the chapter

http://dx.doi.org/10.5772/67236

Abstract

This chapter provides a review on the state of art of the use of the visible near-infrared (vis-NIR) spectroscopy technique to determine mineral nutrients, organic compounds, and other physical and chemical characteristics in samples from agricultural systems—such as plant tissues, soils, fruits, cocomposted sewage sludge and wastes, cereals, and forage and silage. Currently, all this information is needed to be able to carry out the appropriate fertilization of crops, to handle agricultural soils, determine the organoleptic characteristics of fruit and vegetable products, discover the characteristics of the various substrates obtained in composting processes, and characterize byproducts from the industrial sector. All this needs a large number of samples that must be analyzed; this is a time-consuming work, leading to high economic costs and, obviously, having a negative environmental impact owing to the production of noxious chemicals during the analyses. Therefore, the development of a fast, environmentally friendly, and cheaper method of analysis like vis-NIR is highly desirable. Our intention here is to introduce the main fundamentals of infrared reflectance spectroscopy, and to show that procedures like calibration and validation of data from vis-NIR spectra must be performed, and describe the parameters most commonly measured in the agricultural sector.

Keywords: vis-NIR spectroscopy, calibration and validation methods, plant mineral analysis, fruit analysis, soil analysis, fruit organoleptic characteristics

1. Introduction

One of the challenges of the twenty-first century is to achieve a more productive agriculture, while improving the safety and quality of food. The food industry has to feed a population that is in continuous increase, bearing in mind that these systems have to respect the environment, should optimize natural resources in each area, and anticipate changes in temperature and rainfall that will occur in the future as a result of climate change. Proper soil management and fertilization of crops will be crucial to increasing the capacity of agriculture, to the provision of products of high added value, and to the protection of crops against pests and diseases. To do this, in each of the steps ranging from the production of fruits and vegetables in the field to the development of industrial products, it is necessary to determine a great number of physical and chemical parameters in the soil, plants, fruits, compost, and byproducts from food processing industries. Currently, the traditional techniques of analysis of such samples are being replaced by spectroscopic techniques—one of which is visible near-infrared spectroscopy (vis-NIRS). This technique has a number of advantages over the traditional methods, as it (i) is a method of nondestructive analysis, (ii) does not pollute the environment, because it does not use chemical reagents, (iii) is cheap and fast, (iv) measures many parameters in a single analysis, and (v) can perform analyses *in situ* and online for a large number of samples per minute.

The aim of this chapter is to provide an updated review of the current state of vis-NIRS as a technique for the estimation of physical and chemical parameters in samples derived from agricultural systems, such as soils, plants, fruit, compost, and products derived from food processing industries. The chapter starts by describing the basic principles of this technique and the different ways in which the equipment can be calibrated, detailing the statistical tools that are useful to establish that the calibration and the estimation of the desired parameters are valid. We will describe the parameters that can be measured by vis-NIRS in samples, with the emphasis on soil, plants, fruit, compost, and byproducts from the industrial sector that processes the output of agricultural systems. A basic explanation of the parameters measured in these samples will be given, together with a description of how they are measured and the mathematical tools used, focusing on the most novel issues.

2. Fundamentals of infrared diffuse reflectance spectroscopy

Spectroscopy in the near infrared or NIRS (near-infrared reflectance spectroscopy) is a tool that has been used widely for the rapid determination of organic components. For example, NIRS readout for nutrient level estimation on citrus leaves, using FT-NIR spectrometer and 64 scans per sample, takes 1–2 min. The only pretreatments of the sample required prior to analysis are drying, crushing, and mixing, in the case of solid matrices. Samples can also be scanned when fresh, as in the work of Huang et al. [28]. All this bestows on this technique several advantages over other, more sophisticated spectroscopic or analytical methods. The operating principle of the NIRS technique requires that the energy absorbed in the near-infrared region by a sample causes covalent bonds of C-H, O-H, and N-H, important components of organic substances, to vibrate in different forms [1]. Within the field of NIRS, two main types of fundamental vibrations are considered: stretching, which involves a change in the length of a bond, and bending, which involves a change in the angle between two bonds. Overtones appear when a vibrational mode is excited at a frequency higher than that of the fundamental vibration.

This infrared fraction comprises wavelengths between 780 and 2500 nm (12,500–4000 cm^{-1}, expressed as a wavenumber, **Table 1**).

Group	**Aliphatic hydrocarbons**	**Aromatic hydrocarbons**	**Carboxylic acid**	**Amines**	**Water**
Frequency range (cm^{-1})	9100–7800 (overtone of CH-stretching)	ca. 9000 (overtone of CH-stretching)	ca. 6900 (overtone of CH-stretching)	7000–6500 (overtone of NH-stretching)	7500–6400 (overtone of OH-stretching)
	7700–6900 (combination)	7300–6900 (combination)	ca. 5250 (overtone of CO-stretching)	5200–4500 (combination)	5400–4900 (combination)
	6300–5500 (overtone of CH-stretching)	ca. 6000 (overtone of CH-stretching)	4900–4600 (combination)		
	5000–4100 (combination)	4700–4000 (combination)			

Table 1. Absorbance signals in the near infrared for the major chemical groups present in organic matter.

There is a relationship, both quantitative and qualitative, between the chemical composition and the spectrum recorded in the near-infrared. Hence, samples having different organic compositions have different infrared spectra. But, interpretation of the spectra is tremendously complex, although the spectral characteristics of each compound are unique, as their amplitudes sometimes overlap.

Before the NIR spectrum of a sample can be used for the determination of a compound or specific element, a calibration for this compound or element must be developed. In an NIRS spectrum, the various constituents of the sample have some overlapping peaks; thus, the measurements made with NIRS must be calibrated with samples of known chemical composition in order to extract the desired information using NIRS [2].

3. Calibration and validations of data from NIRS

Chemometrics includes all methods of multivariate calibration in the field of analytical chemistry. Unlike univariate calibration, where a spectral peak (height or area) is correlated with the reference concentration, multivariate calibration uses the entire spectrum structure with a large amount of spectral information to correlate with the reference concentration.

The establishment of a model for the use of NIRS data in the analysis of samples consists of the following steps: (1) introduction of the spectral and concentration data; (2) preprocessing of the spectral data; (3) definition of the appropriate frequency range; (4) validation and optimization of the method; (5) definitive calibration; and (6) routine analysis.

(1) Introduction of the spectral and concentration data. This process begins with the selection of the group of samples for calibration, which must be well defined statistically, and pretreatment of the samples to assess measurement errors. The dispersion of incident radiation, also known as the scatter effect, produces a low selectivity (quality of being able to tune in to one particular frequency while blocking out other unwanted frequencies) of the NIR spectral information [3]. This is due to physical phenomena—such as the texture, size, and geometry of the particles that make up the sample [4, 5]—and to changes in the refractive index of the material which interacts with the radiation, causing numerous unwanted variations in the NIR spectral data [6–9]. Depending on the complexity of the samples, between 20 and 200 samples are necessary to develop a multivariate calibration method. The greater the number of samples, the more representative is the calibrations achieved. The samples should have a normal distribution, cover the entire range of concentrations of the parameters that are to be estimated by NIRS, and should not have areas where uncertainty is high and errors can be significant. For instance, the NIR spectrum of water (transmission measurement, optical path length: 2 mm) shows a total absorption between 5200 and 4000 cm^{-1} and below 4000 cm^{-1} a strong contribution of spectral noise. Finally, for each sample a classical analysis of the desired components is carried out, to obtain the so-called reference values, and its NIR spectrum is obtained.

(2) Preprocessing of the spectral data. The spectral pretreatment that improves the signal/noise ratio must be chosen. For example, the problems of baseline displacement need to be eliminated. The procedures for preprocessing of the NIRS spectrum, to obtain a good correlation between the spectral data and the concentration, include: no data preprocessing (NDP), first derivate (FD), application: it is used to emphasize pronounced, but small features compared to enormous broad-banded structures or on the evaluation of broad bands that get a steeper shape, so it can be evaluated more easily; second derivate (SED), application: similar to first derivative, where even extremely flat structures can be evaluated, but the spectral noise is enhanced as well, the most widely used methods here are the Savitzky-Golay [10] and Norris [11] methods; standard normal variate (SNV [6]; multiplicative scatter correction (MSC [3, 12]), application: it is used for measurements in diffuse reflection; detrending (DT), which is usually applied in conjunction with SNV; spectral smoothing (SS), for which the most used are the Savitzky-Golay [10] and Fourier transformation [13] or vector normalization (VN), application: in a measurement in diffuse reflection, the interfering influences of different material densities or particles sizes can often be minimized; maximum-minimum normalization (MMN), application: similar to vector normalization; subtraction of a straight line (SSL), application: a linear tilt of the baseline shift is eliminated; linear offset subtraction (LOS), application: linear baseline shifts are eliminated. The optimum method depends on the system to be analyzed. Generally, SSL, VN, or FD leads to better calibration.

(3) Definition of the appropriate frequency range. Once the calibration samples have been selected and then analyzed by the reference method and NIRS, a correlation between the spectral and analytical data is searched for [14]. For this purpose different statistical treatments are used, such as multiple linear regression (MLR [15]), principal components regression (PCR), and partial least squares regression (PLSR) as linear methods and use of artificial neural networks (ANN) as a nonlinear method. PLSR is the one most commonly used

[16, 17]. The best correlated frequency ranges are assessed and then selected based on the coefficient of determination, R^2, and a corresponding low error of analysis (*root mean square error of cross validation/root mean square error of prediction* (RMSECV/RMSEP), see equations (4) and (5)). Typically, an R^2 value of 0.75–1.0 indicates an acceptable correlation and it depends on the type of sample. Good values for R^2 are larger than 0.90 for solids and larger than 0.99 liquid measurements. The total absorption of water (frequency range 5200–4000 cm^{-1}) yields relative absorbance values (A) greater than 2.5. The use of dried samples prevents interference of water in the aforementioned frequency range. Spectral noise is usually found below 4000 cm^{-1} and gives relative absorbance values lower than 0.7. Thus, this region should not be included to establish a calibration. Values of A between 0.7 and 1.0 generally give better results. Besides, modern FT-spectrometers allow the use of absorbance values of up to 2.5 for the calibration.

(4) Validation and optimization of the method. To choose the best calibration for the regression equation with linear models (PLS algorithm, for instance), the instrumental software combines different methods of data pretreatment and frequency ranges. Then, it provides as output the corresponding mean error of prediction and R^2 for a given number of factors. The quality of the calibration is evaluated by the validation, which consists of comparing the concentrations predicted by the calibration with the reference values of samples not used in this calibration [18]. There are two types of validation: internal or cross validation and external validation. In internal validation, a sample, or group of samples, is taken from the set of samples. With the calibration obtained using the remaining samples, the concentrations in the previously separated samples are predicted. The samples are interchanged until all have been used once for the validation. In external validation, all samples are used for calibration and prediction is performed for additional samples [19]. Since optimal frequency windows and pretreatments of signals cannot be anticipated, they are generally determined empirically by trial and error. These values are calculated for a growing number of factors. The concentration and spectral data are encoded in matrix form and reduced to a small number of factors called "rank." To some extent, the factors or principal components are "information units," as may be the case for the concentration of a sample component. In many cases, there are several combinations of frequency window and pretreatment of spectral data of comparable quality for the prediction of analytical results. In these cases the combination that has fewer factors is recommended, as it generally will be more stable (**Table 2**). The optimum method is number 2 (mean error of prediction 0.07% and optimum rank 6). However, it is possible to manually set a lower rank in order to get a better result.

Number	Data preprocessing	Frequency ranges (cm^{-1})	Optimum rank	Coefficient of determination (R^2)	Mean error of prediction (%)
1	NDP	9000–5200	9	0.998	0.16
2	SSL	9000–5200	6	>0.999	0.07
3	VN	9000–5200	8	0.996	0.42
4	SSL	7000–5200	8	>0.999	0.07

No data preprocessing (NDP), subtraction of a straight line (SSL), and vector normalization (VN).

Table 2. Method optimization using the PLS algorithm for NIRS analysis of CH_3OH concentration in a mixture of CH_3OH, C_2H_5OH, and C_3H_7OH.

There will be anomalous samples and spectra, called "outliers," which affect the predictive ability of the NIRS equations obtained [20]. It is important not to remove them until one has a clear explanation; to make this decision one has to take into account the *t*-test ("Student's *t*-test" Eq. 1) and H (Mahalanobis distance, Eqs. 2 and 3) values, among others.

t-Test applied to each wavelength gives an idea about the weight of each wavelength in the calibration. The higher the value in the *t*-test, the more important it is. If it is higher than 10, it is considered essential to take part in the calibration equation.

$$t = \frac{Y_{pred} - Y_{ref}}{SEC * \sqrt{1-H}} \tag{1}$$

SEC is the standard error of the calibration, and *H* is the spectral error. In NIRS, $t > 2.5$ are considered significative and kept in the calibration.

Chemical outliers can be recognized after applying a *t*-test since they present significative differences between the composition value provided by the reference method and the regression model.

To detect spectral outliers, the Mahalanobis distance is particularly useful. For MLR models it is calculated as follows:

$$H = \frac{K}{n} \tag{2}$$

For the models PLSR and PCR, it is expressed with the following equation:

$$H = \frac{K+1}{n} \tag{3}$$

n is the number of spectra in the dataset and *K* is the number of selected wavelengths. $H < 3$ means that the samples belong to the population.

The statistics used in the evaluation, selection, and validation of the calibration equations are as follows:

- *Determination coefficient of the calibration* (R^2_c)/*Determination coefficient of the cross validation* (R^2_v). This establishes a correlation between the analytical data obtained in the laboratory and those predicted by the calibration equations for each of the components analyzed. As mentioned above, an R^2 value of 0.75–1.0 indicates an acceptable correlation. Some calibrations with an R^2 value <0.75 may be useful for monitoring purposes. Thus, an R^2 value of 0.50–0.69 distinguishes between low, medium, and high values; an R^2 of 0.30–0.049 distinguishes between low and high values; and with $R^2 < 0.29$, it is better not to analyze [21, 22].

- *Root mean square error of estimation* (RMSEE). This is the error associated with the differences between the analyses performed in the laboratory using the reference methods and the results of the analysis by NIRS technology, for each of the parameters determined in the samples used in the calibration. This value of this statistical parameter should be as low as possible. It is calculated using the formula:

$$RMSEE = \sqrt{\frac{1}{M-R-1} SSE} \tag{4}$$

where M is the number of samples in the calibration set, R is the number of principal components (factors), and SSE is the squared sum of the differences between the actual and estimated values. It is preferable to compare this type of error with the error that can occur with traditional methods of analysis and decide whether the error is acceptable for routine use. The prediction error (P) is the accumulation of the errors of the reference concentrations (R), of the NIRS data, and of the calibration. The lower the ratio of the errors (P/R), the greater the accuracy of the NIRS model obtained: $P/R = 1–1.5$, excellent; $P/R = 2–3$, good; $P/R = 4$, moderate; and $P/R = 5$, poor [23].

- *Root mean square error of cross validation/Root mean square error of prediction* (RMSECV/RMSEP). Following calibration the cross validation error is obtained. This error is the one that should be taken into account most closely when evaluating the calibration. To calculate it, considering the number of samples in the set and the differences between the estimated values and those obtained by standard methods of analysis, the following formula is used:

$$RMSECV = \sqrt{\frac{1}{M} \cdot \sum_{i=1}^{M}(Differ_i)^2} \tag{5}$$

- *Residual prediction deviation* (RPD). This is defined as the ratio between the standard deviation of the reference data and the RMSEE/RMSECV. One researcher [24] provided a guide to evaluate calibrations performed with environmental samples, based on the R^2 and the RPD, as follows: excellent, $R^2 > 0.95$ and RPD > 4; good, $R^2 = 0.9–0.95$ and RPD = 3–4; quite good, $R^2 = 0.8–0.9$ and RPD = 2.25–3; quite useful, $R^2 = 0.7–0.8$ and RPD = 1.75–2.25. RPD is of the same significance as R^2 explained variance. The R^2 also allows a qualitative evaluation of the error of prediction during the validation process.

- *Bias*. This is the difference between the mean value predicted by FT-NIRS and the mean value of the reference predictive model and the residual prediction deviation (RPD, [20, 25, 26]): M is the number of samples used in the calibration, xi is the result obtained by NIRS, and yi is the result obtained by the reference method for sample i:

$$Bias = \frac{1}{M} \sum_{i=1}^{i=M} (xi - yi) \tag{6}$$

In the presence of laborious and troublesome datasets, it is possible to ask for high-performance external NIR calibration services such as those provided by private companies to optimize and validate the method.

(5) Definitive calibration. After all the "outliers" have been eliminated and the optimal parameters determined (for example, R^2, RMSEE, and RPD for the calibration and R^2, RMSEP, RPD, and bias for the validation), the final calibration model is evaluated for the analysis of new samples.

(6) Routine analysis. Here the optimum chemometric model is used to analyze quickly unknown samples. The Mahalanobis distance can alert one to samples ("outliers") that are outside the calibration range or do not fit the model well.

4. Mineral nutrients and organic compounds in different samples from agricultural systems

4.1. Plants

Fruit and vegetable crops, in order to achieve good vegetative growth and maximum production with good quality fruit, require a good nutritional status, maintaining a proper balance of nitrogen, potassium, phosphorus, and trace elements such as manganese, boron, copper, and magnesium. The guide to the nutritional status of crops is based on the method known as "sufficiency range" (SR), which establishes—for each nutrient—the ranges of values considered to be normal and to represent deficiency or toxicity [27]; or it can be based on the establishment of indices of dependent nutrients, in which each index includes two or more nutrients, the so-called "Integrated System of Diagnosis and Recommendation (DRIS)." But, a good fertilization program also should pay attention to the changes that occur in the mineral status of plants in their different phenological stages so that the application of fertilizers can be adapted to the requirements of the plants at all times. To obtain this information, it is necessary to perform mineral analyses of leaves by ICP-OES or AA after acid digestion of the samples, as well as analysis of the C/N ratio, which involves the analysis of a large number of samples with all that this entails. Currently, the available knowledge of reflectance spectroscopy in the near-infrared (NIR) part of the spectrum can be used to determine the nutritional status of crops quickly and cheaply. The mineral composition of an organic matrix can be estimated by NIRS, from the spectra in the range 700–2500 nm, due to the association between the minerals and the organic functional groups or the organic matrix itself [28]. There are no infrared absorption bonds in the mineral species of macro- and micronutrients, but NIRS determines bonds within organic compounds that are negatively related to inorganic materials. If mineral matter is bound to organic compounds, the distortion of the spectrum is detectable at certain wavelengths, suggesting that NIRS can quantify inorganic materials using their ratio to the organic matter [29].

Numerous studies show that the NIRS technique, together with multivariate analysis and partial least squares regression (PLSR), provides a powerful tool for the interpretation and analysis of spectra. For example, NIRS technology has been used successfully to predict the nutritional status of leaves of apple [30], alfalfa [31], sugar cane [32], root crops [33], yerba mate [34], and citrus [35, 36]. It has been observed in citrus leaves of different varieties including lemon, mandarin, orange, and grapefruit—high accuracy regarding the estimation of N ($R = 0.99$) and Ca ($R = 0.98$) as well as acceptable estimates for K, Mg, Fe, and Zn [37]. However, good calibrations for the estimation of P, B, Cu, and Mn were not obtained. Furthermore, the concentrations of nutrients could be estimated with a single calibration model, regardless of the variety of citrus analyzed. In yerba mate plants, the prediction was good for P and Cu but not for K, Ca, Na, Mn, or Zn [34]. These data show that the NIR spectral response depends on the species studied, so for each species it is necessary to make the appropriate calibrations—but these are valid for different cultivars of the same species.

4.2. Soils

Soil is a natural resource that is vital in agriculture for the production of food, fiber, and energy; but it serves also as a platform for human activities, constitutes an element of the landscape, is an archive of cultural heritage, and plays a central role as a habitat and gene pool. It stores, filters, and transforms many substances, including water, nutrients, and carbon (C). In fact, it is the largest C "store" in the world (1500 gigatonnes). All these functions must be protected because of their socioeconomic and environmental importance.

Fundamentally, the soil is a complex matrix of organic matter, minerals, water, air, and microorganisms. The soil organic matter is only a small part, but it plays a big role in both the physical and chemical properties of soil as well as in the development of crops. This soil fraction comprises humus (material that is decomposed, dark, and colloidal in nature) and materials such as the roots and aerial parts of plants and the bodies of insects and other animals that are deposited on the ground. The content of organic matter normally found in the soil is small, only about 1–5% by weight, of which 85–90% is humus and only a small part is the nonhumified remains. The mineral phase is a mixture of materials that differ in their composition and properties. Typically, this fraction is characterized by particle size. Stones, gravel, and sand represent the coarse fraction, while smaller particles like silt and clay constitute the fine fraction of the soil. The clays also can be classified according to the negative charges on their surfaces—some minerals are more negative than others—and this property also influences the chemical characteristics of the soil. There is a parameter that is very useful for measuring these chemical properties, namely, the cation exchange capacity (CEC)—defined as the maximum amount of cations that a soil can fix. Water is another significant fraction of the soil, and its content depends on the amount and size of the pores in the soil. In plants, water is the major constituent of protoplasm (85–95%) and is essential for physiological processes such as photosynthesis, nutrient transport, and maintenance of turgor. The air is another important fraction of the soil. Its oxygen is essential for the respiration of roots and microorganisms. When aeration is poor, organic matter is oxidized slowly, the activity of aerobic microorganisms is paralyzed, and only anaerobes are active, giving rise to reduced forms of elements that are usually toxic to plants. The most common microorganisms in soils are nematodes, protozoa, and rotifers—whose activities also determine the physiochemical soil characteristics, as they have the ability to degrade highly resistant organic compounds such as cellulose and lignin, and can even degrade minerals, thus releasing plant nutrients.

The study of soil should take into account the different phases of the soil and must be directed toward two main objectives: (i) consideration of its various properties, with special emphasis on plant productivity (that is, practical or applied aspects); and (ii) scientific, especially chemical, study—to determine the variation of productivity and find ways for soil conservation and improvement. In recent years, it has been observed that NIRS (using visible-near infrared) can be very useful for characterization of soils. This technique has many advantages: sample preparation is easy as it only requires the drying and grinding of the soil, reagents nontoxic to the environment are used, measurements are made in a few seconds, a single scan can show multiple properties, and the technique can be used both in the laboratory and *in situ*. The parameters that can be measured in the soil using NIRS are described below.

4.2.1. Soil organic matter (SOM)

In the laboratory, it is difficult to separate the organic and inorganic material of a soil, so an estimate of the content of organic matter is obtained indirectly through the analysis of an element that is a constituent of all organic substances, namely C. Once the amount of organic C present in a soil sample is known, the percentage of organic matter with respect to the total weight of the soil can be estimated indirectly. Classically, it has been determined by various methods—such as calcination of the soil sample, oxidation with potassium dichromate, or oxidation with hydrogen peroxide. With regard to NIRS, a large number of studies have shown that this analytical technique is very useful for estimating SOM. For this, absorption bands in the NIR region which result from the stretching and bending of NH, CO, and CH groups, that form part of the organic material, are used [38].

The 1990s saw the identification of the absorption bands of wavelength (nm) 1100, 1600, 1700–1800, 2000, and 2200–2400 as the most useful for measuring organic C. Since 2000, the technique of NIRS has been perfected and adapted to the soil and climatic conditions of each area. Thus, in Australian soil, it has been also observed that if the absorption spectrum was made in the vis-NIR region—that is, including the visible region—better results were obtained than with NIR alone [39]. Another problem that researchers have faced in achieving good calibrations has been that the spectral response can change depending on the mineral fraction of the soil, composition of organic matter, texture, and soil moisture content [40–42]. All these problems have been solved by optimizing the way of taking samples (local and regional scales) [43, 44], choosing the most appropriate mathematical models for the calibration [45–48], establishing covariance models, or eliminating certain factors that make the model weak [49, 50]. For example, in saline soils of El-Tina Plain (Egypt), it has been compared several regression techniques to estimate the organic matter content of soils [51]. Specifically, they used PLSR, support vector regression (SVR), and multivariate adaptive regression splines (MARS) and found that the best calibration was obtained with MARS with continuum removed reflectance preprocessing (R^2 and RMSE were 0.89 and 0.19, respectively). The calibration model to estimate the organic matter in an area of the Grand-Duchy of Luxembourg have been improved, taking into account the amount of water in the soil [52]; it was also considered by NIRS, using the reflectance values at 1800 and 2119 nm and calculating the normalized soil moisture index (NMSI).

4.2.2. Soil mineralogy

The mineral fraction of the soil occupies almost half of the soil volume. Its composition and concentration as well as the proportion of different minerals determine important properties such as texture, structure, and CEC. These properties also determine other soil characteristics such as the availability of nutrients to agricultural crops. Classical methods for the determination of clay minerals are qualitative and are based on XRD (XTR). However, some researchers [53, 54] made the first tests to see if NIRS could be used to estimate soil minerals; later, these same authors [55] compared the NIRS results with XRD analysis, concluding that NIRS is an effective method to determine the mineralogy of the soil. This is because soil minerals absorb light in the UV, visible, vis-NIR, and mid-NIR parts of the electromagnetic spectrum. For example, Fe oxides absorb UV radiation while phyllosilicates (clay minerals) have varied spectra in the vis-NIR. Overall, this technique has been used to estimate the Fe oxides goethite (α-FeOOH) and hematite (α-Fe2O3), clays of the kaolinite, illite, and smectite types, and carbonates [56, 57].

4.2.3. *Soil texture and CEC*

The water dynamics and aeration of a soil depend on its structure and texture, and these parameters are important for the development of both plants and microorganisms, so they need to be evaluated. These parameters also determine the leaching of fertilizers and pesticides in agricultural soils. Generally, soil texture is defined as the ratio (in percentage by weight) of particles smaller than 2 mm in diameter and classified as sand (2–0.02 μm), silt (0.02–0.002 μm), or clay (0.002 μm). Ben-Dor and Banin [58] found that the clay content may be estimated by analyzing the absorption bands of O-H in water, and those of Mg-, Al-, and Fe-OH in the mineral fraction of the soil. Curcio [59] used visible and near-infrared (VNIR, 400–1200 nm) and shortwave infrared (SWIR, 1200–2500 nm) reflectance domains to estimate soil texture in three agricultural areas of Italy (Bompensiere, Dirillo, and Pietranera), and obtained a good calibration by using the PLSR method, the accuracy being good for the clay fraction (RMSE = 5.8%, R^2=0.87) and satisfactory for sand (RMSE = 7.7%, R^2=0.80) and silt (RMSE = 7.2%, R^2=0.60).

The CEC is traditionally measured by the method of Chapman [60], based on saturating the soil with sodium. However, the vis-NIR technique can also estimate this parameter in soils accurately, if methods of multivariate regression are used instead of simple bivariate relationships, and it is suitable for measurements of peak intensities in the mid-IR range. Recently, Ulusoy et al. [61] obtained a good prediction of CEC using an analysis of PLSR, both in the laboratory and for online measurements in the field—although the calibration was much better for the data obtained in the laboratory.

4.2.4. *Plant nutrients*

Due to the importance of the mineral nutrition of plants in the yield and quality of fruit and vegetables, one of the most common practices in agriculture is the analysis of the soil content of N, P, K, Fe, Ca, and Mg. This information is particularly important when optimizing fertilization programs. In most farming systems N is the element most commonly applied, followed by K, P, Ca, and micronutrients. These nutrients do not have a specific absorption spectrum in the vis-NIR region. Generally, the correlations between the "real" concentrations of these nutrients and those estimated by NIR are highly variable, the variability coefficients (R^2) being in the following range: N (0.11–0.55), available potassium (0.56–0.83), exchangeable potassium (0.11–0.55), Ca (0.75–0.89), Fe (0.64–0.91), Na (0.09–0.44), Mg (0.53–0.82), and P (0.23–0.92) [40]. This variability may depend on many factors, so local-scale calibrations are recommended to achieve greater accuracy.

4.2.5. *pH*

Soil pH is an important regulator of fertilization. The solubility of the nutrients is pH-dependent, and plants may decrease nutrient uptake if the soil pH is not suitable. Other pH-dependent factors include the biological activity, decomposition of organic matter, and mineralization. Generally, the ideal soil pH for plants is between 5.5 and 6.5. Soil pH, or more specifically the H^+ ions, has no direct response to NIR but its value can be estimated well with this technique if the appropriate covariations are applied to components that do exhibit activity in the NIR, such as organic matter and clays [62]. In different experiments it has been found that pH calibration gives R^2 values between 0.55 and 0.77 and an RMSE of 0.3–0.5 pH units. These parameters could be further improved if specific calibrations were made at the local scale, while studying in detail which covariance parameters have a direct influence on the NIR and thus should be used.

4.2.6. *Heavy metals and other soil contaminants*

Heavy metals are potential pollutants of air, water, and soil and of plants when taken up in sufficiently high amounts; this pollution will also affect other links in the food chain. In most agricultural soils, there are small amounts of As, B, Cd, Co, Cr, Cu, Mo, Mn, Ni, Se, and Zn, but when normal values are exceeded this can cause soil pollution and phytotoxicity, negatively impacting the agronomic performance. Usually, heavy metals are measured by atomic absorption or ICP. In the vis-NIR region these metals do not absorb energy, but their concentrations can be estimated if used as covariates with other components that do possess absorption spectra [40]. For example, they can be related to the organic matter, hydroxides, sulfides, carbonates, oxides, clay minerals, or soil texture (Stenberg et al. [40]). Todorova et al. [63] investigated the use of NIRS to estimate the concentration of heavy metals (Zn, Cu, Pb, Cr, and Ni) in various soils of Stara (Zagora Region, Bulgaria), using the PLS type of calibration. The best validation of the method was observed with Cu, while it allowed estimation of whether the concentrations of Zn, Pb, and Ni were low or high; however, Cr gave the weakest validation. These authors also noted that as the number of samples in the validation process increased, the RMESP values decreased.

In soils, hydrocarbons can also be measured with the vis-NIR technique, to establish the degree of contamination of soils that have suffered spills of petroleum products. Okparanma [64] used this technique to make soil maps in which the concentration of polycyclic aromatic hydrocarbons and their equivalent toxic concentrations in soil from Niger (Nigeria) were indicated. The data of this study revealed that the elaboration of soil spectra between 300 and 2500 nm, together with a PLS calibration, permitted the estimation of the concentration of hydrocarbons without significant differences from the results obtained by the conventional method of gas chromatography-mass spectrometry.

4.2.7. *Soil moisture*

There are a multitude of reasons to measure the water content in agricultural soils because water is fundamental to the development of plants and for soil biology, besides regulating important processes in the soil such as nitrification/denitrification, leaching, and erosion. In the laboratory, although there are many methods of measurement, the traditional one is based on weighing the sample fresh and then after drying, calculating the percentage of water relative to the dry weight of soil. Water produces an absorption spectrum in the NIR due to expansions and stresses of O-H bonds. Water incorporated into the soil in clay minerals absorbs at wavelengths around 1400 and 1900 nm. The main problem researchers have had in the calibration of this parameter is due to the fact that the water found on the surface of the minerals in thin layers and in the pores tends to decrease the albedo (the percentage of radiation that the soil reflects), changing the refractive index. As the porosity and refractive index of soil particles vary between soil types, a relationship between the albedo and water content cannot be given. However, some authors have successfully used a multivariate calibration with data from the NIR spectral bands to estimate the water content. For example, Bullock et al. [65] found a good correlation using a PLSR of the regions of 1100–2500 nm, and Ben-Dor and Banin [58] produced a good regression for samples having a water content of 0.2–11.6% using an MLR (multiple linear regression) calibration. The problem with all these calibrations appears when one wants to have a single calibration for soils of very different geological origins; thus, calibrations at the local scale are recommended [62, 66].

4.3. Fruits

Fruits and vegetables from agricultural plantations must maintain their optimum quality, whether they are destined for fresh consumption or for processing. As quality rises so do prices, so it is necessary to determine the intrinsic characteristics and external appearance of the fruits and vegetables. This information can be used to exclude fruits of poor quality, and can be provided to the consumer/industry to inform them of the added value of the product that is being offered. External defects—such as bruises, injuries from cold and wind, cracks in the skin, and contamination by pathogens—cause significant economic losses. Parameters such as total soluble solids content (TSS), acidity, and water content, which are related to the flavor and aroma of fruit and vegetables, serve to define their organoleptic quality. Therefore, currently, analytical methods are being developed that allow accurate, fast, and noninvasive determination of the qualities of agricultural products. In the case of the appearance of the fruits, computer vision technology—which integrates data acquisition, processing, and analysis—has great potential for the automatic inspection of the appearance of the products. The internal quality of the fruits can be estimated accurately by spectroscopy in the visible and infrared (vis-NIR), because most of the organoleptic characteristics are related to functional groups of the type C-H, N-H, and O-H. What follows is a brief summary of the most significant parameters that can be measured with these spectrometric techniques.

4.3.1. Total soluble solids

This parameter is used to measure the approximate amount of sugars in fruit juices, wine, or liquids processed in the agri-food industry, and is used to track *in situ* the evolution of ripening of the fruits and their optimal harvest time. The determination is made by refractometry and expressed in Brix, equivalent to grams of sugar per 100 ml of juice. Using the vis-NIR technique with a calibration model based on PLS, many researchers have obtained high correlation coefficients (R) (0.80–0.95) for fruits of mango, strawberry, apple, table grape, banana, blueberry, and bell pepper [67]. Recently, it has been shown that this technique can measure online the TSS of pear fruits (five fruits per second) as they pass along a transport chain [68]. The authors noted that the relationship between the wavelengths of 681 and 822 nm and a PLS calibration model allowed estimation of both the TSS and the healthy pears that had no apparent surface damage; therefore, they suggested that this technique could be integrated into industrial processes to select good quality fruit quickly, thereby reducing the labor required for both processing and laboratory analysis. Also, with portable vis-NIR equipment, the state of maturation of the grape variety Sangiovese could be determined *in situ*, by calculating the index of absorbance difference (IAD) from the values at wavelengths of 560 and 640 nm. These values correlate with TSS as well as with parameters such as titratable acidity (TA), firmness (DI), and anthocyanins—which allows one to know quickly and accurately the date on which the fruit should be harvested [69].

4.3.2. Total titratable acidity

This parameter describes the total concentration of acids in food, vegetables, or fruit. It is determined by an acid-base titration (soluble acids determined as citric, malic, lactic, oxaloacetic, succinic, glyceric, phosphoric, hydrochloric, fumaric, galacturonic, glyceric, tartaric acids, etc.). Acidity influences the taste of food (roughness), the color, the microbial stability, and the quality of conservation, and is determined by an acid-base titration using 0.1 N NaOH as the base and phenolphthalein as the indicator. Estimations by NIRS of this parameter and the pH of the fruit are as good as those found for the TSS. Thus, with the corresponding calibration data obtained by NIR, values of R between 0.80 and 0.82 have been observed in Chinese bayberry, apple, strawberry, table grape, and grape using wavelengths between 320 and 1650 nm [67]. In Spain, acidity has been studied using online NIRS combined with chemometric techniques (PCA, LDA, and PLSR) in fruits of different olive varieties; this gave good estimates of the free acidity ($R^2 = 0.83$), water content ($R^2 = 0.76$), and fatty acid content ($R^2 = 0.83$). For the calibration, a reflectance spectrum of intact olive fruits in the wavelength range 1000–2300 nm was produced and then samples of a paste prepared from these fruits were analyzed in the traditional manner. The estimation of these parameters improved when the calibration was performed for each stage of ripeness [70].

4.3.3. Contents of water and dry matter

For the food industry, the moisture content is an important quality factor of fruit and vegetables, whether fresh or processed, and influences their conservation and deterioration. The dry matter content is obviously very important when calculating the contents of other constituents of fruit and vegetables on the basis of the dry matter, which is uniform and less variable than the fresh weight. Water is the major component of all fruit and vegetables, representing between 60 and 96% by weight. The methods used most commonly for its determination are drying methods; the percentage water content is calculated as the loss in weight due to elimination by heating under standard conditions. Pu et al. [67] stressed that the vis-NIR technique is useful for measuring the water content in fruits of mango, strawberry, mushroom, banana, and soybean using spectra in the range of 400–1000 nm.

4.4. Grains and seeds of cereals, grasses species, and legumes

Cereals are important in animal feed not only because they provide energy but also since they contribute 70% of the protein in the diet and are especially important for pig nutrition. Similarly, soya is a major source of vegetable protein in animal feed formulations. It is important that the animal diets have a proper balance regarding the contents of amino acids.

The main nutritional parameters in the grains and seeds of these crop plants determined by the NIRS technique with regard to animal feed are the moisture and protein contents, representing the biological value. Other parameters analyzed are the contents of lipids, carbohydrates, and ash.

4.4.1. Amino acids

In 1978, Rubenthaler and Bruinsma [71] developed the first calibration equations for the determination of lysine in wheat and barley. Subsequently, Fontaine et al. [72] determined the total contents of methionine (Met), cysteine (Cys), lysine (Lys), threonine (Thr), tryptophan (Trp), and other essential amino acids in a population of cereal and sorghum samples. The spectra were first treated with SNV (recommended for samples with <15% moisture) and trend to reduce differences in the spectra that are caused by particle size effects only and not by changes in the constituents. In this way, the validation of the calibration equations showed that 70–98% of the variance of the amino acids in the samples could be explained using the NIRS technique, especially for Lys and Met—the amino acids most limiting to animal nutrition. Also, Kovalenko et al. [73], in the analysis of soybean samples, applied the MSC mathematical treatment (let to remove background spectroscopy) to the spectral data together with the PLSR regression model and obtained a determination coefficient (r^2) of 0.91 for Lys. However, the concentrations of Cys and Trp did not exhibit a good correlation with the spectral information, the r^2 value for Trp being 0.04.

4.4.2. Other organic matter

With respect to dry matter, lipids, total protein, carbohydrate, and ash, Ferreira et al. [74] and Wang et al. [75] established models to determine the protein and lipid contents in both soya and fava beans. These authors obtained high R^2 values for protein (0.81 vs. 0.94); however, for lipids the values were slightly lower (0.71 vs. 0.66). Both groups used as a mathematical treatments: standard normal variate transformation (SNV; let to correct scattering effects caused by physical differences between samples) and first derivate. For the rest of the components, Ferreira et al. [74] obtained calibrations giving high predictability for dry matter, ash, and carbohydrates (RMSEP of 0.38–3.71%), the prediction being poorest for carbohydrates (R^2_c = 0.50 and RPD_c = 0.83). Wang et al. [75] found RPD values of 2.95 and 2.50 for starch and total polyphenol content, respectively.

4.4.3. Toxic substances

Some seeds may contain substances that are antinutritional or toxic in nature, such as L-canavanine in seeds of one-flowered vetch. This is a toxic nonprotein amino acid that can cause a reduction in food intake, particularly in nonruminant animals [76]. These authors determined the content of L-canavanine in one-flowered vetch seeds by NIRS: the calibration equation obtained showed a correlation (r^2) of 0.72, which, according to these authors, only served to separate the samples into groups of low, medium, and high L-canavanine content. However, the equation obtained for the total protein was able to predict with an accuracy similar to that of the reference method, showing a correlation (r^2) of 0.95. Berardo et al. [77] studied the rapid detection of mycotoxins, mainly produced by the fungus *Fusarium verticillioides*, in maize samples. The best predictive ability for the overall rate of infection and *F. verticillioides* was obtained using MPLSR in samples consisting of grains of maize ($r^2 = 0.75$ and SECV = 7.43) and in samples of maize flour ($r^2 = 0.79$ and SECV = 10.95). These authors before the development of the calibration equations applied MSC (multiplicative scatter correction) to remove additive multiplicative effects in spectroscopic data to prevent them from dominating the information signal in the data.

4.5. Forage and silage

The production systems of ruminants are based on forage resources. These forages and—more particularly—maize, wheat, and alfalfa can be conserved as silage. Therefore, quick and reliable knowledge of the quality of forage and silage is very important for technicians and producers. The quality of silage depends—on the one hand—on its nutritional value, which is directly linked to its chemical composition (fiber fractions, nitrogenous materials, minerals, carbohydrates), and—on the other hand—on the quality of its conservation, which is defined by the end-products (lactic, acetic, and butyric acids, ammoniacal nitrogen, soluble nitrogen, etc.) of the fermentation processes. Currently, the main constituents determined by the NIRS technique in forage and silage, and which are important in the feeding of ruminants, are the total protein and protein fractions, soluble and structural carbohydrates, and digestibility of the forage, the latter depending on the content of structural carbohydrates.

Thus, Volkers et al. [78] established calibration equations for samples from different parts of a forage maize crop to predict the crude protein content, obtaining coefficients of determination (R^2) of 0.86–0.96, except for samples of the cobs—which had an R^2 of 0.56. For net energy, the prediction was good—with an R^2 of 0.93 and 0.84 for the entire plant without the cobs and stalks, respectively.

With respect to the nitrogenous fractions, the nonprotein nitrogen/total nitrogen (NPN/TN) ratio in silage is very important for animal nutrition, since it indicates whether the silage has overheated; if the value is greater than 12%, it is considered to have occurred [79]. The acid detergent insoluble nitrogen (ADIN) is the fraction of the total nitrogen that is bound to the cell wall lignin of the plant. This component has low bioavailability but indicates the quality of the silage, as its abundance increases with overheating. Normally, it is not routinely analyzed due to the slowness of the official methods. Consequently, Hermida et al. [79] developed calibration equations by MPLS using first and second derivatives with smoothing average, which led to the removal of spectral noise that makes it difficult to extract relevant information, for determining the TN, soluble nitrogen (SN), NPN, and ADIN in different samples of grass silage. The R^2 values obtained in samples belonging to the validation set were 0.94, 0.92, 0.90, and 0.48 for TN, SN, NPN, and ADIN, respectively. The latter value indicates that NIRS is an acceptable method for the semiquantitative determination of the ADIN fraction.

Nie et al. [80] established calibration equations to predict the total crude protein (CP), true protein (TCP), neutral detergent insoluble protein (NDFCP), and acid detergent insoluble protein (ADFCP) contents in samples of alfalfa. For CP the statistical parameters were $R^2_p = 0.96$ and $RPD_p = 5.07$; for TCP they were $R^2_p = 0.91$ and $RPD_p = 3.31$. However, for NDFCP ($R^2_p = 0.75$, $RPD_p = 1.98$) and ADFCP ($R^2_p = 0.83$, $RPD_p = 2.42$), the prediction was less precise. With these results, the NIRS technique was able to simplify the long and tedious process that determination in the laboratory entails, and predict quickly and empirically the degradability of the alfalfa protein in the rumen; also, these results could be extrapolated to proteins from other forage.

With respect to carbohydrates, Nousiainen et al. [81] established calibration equations to predict neutral detergent fiber (NDF), indigestible neutral detergent fiber (INDF), and digestible neutral detergent fiber (DNDF). For the development of the equations the authors used the MPLSR model and the mathematical treatments: standard normal variate transformation, detrending (SNV-D), and first-order derivatization. The SNV transformation removed scatter effects from spectral data, and corrected scattering effects caused by physical differences between samples. In these parameters, a scatter correction with the standard normal variate transformation combined with detrend eliminated background spectroscopy. The statistics obtained in the cross validation were R^2_{cv} ranging from 0.82 to 0.91 and an RPD_{cv} between 2.39 and 3.33. These authors concluded that the NIRS technique has great potential to predict INDF in grass silage.

Cozzolino et al. [82] developed equations to predict the organic matter, dry matter (DM), acid detergent fiber (ADF), NDF, CP, pH, and *in vitro* organic matter digestibility (DOM) in samples of ensiled whole plants of maize, using second derivative with SNV-D and MSC, which eliminated background spectroscopy. The best statistics obtained in the cross validation were for DM, CP, and ADF, with R^2 values of 0.85, 0.91, and 0.86, respectively. However, for DOM, NDF, and pH the R^2 values showed poor predictive ability, being 0.53, 0.60, and 0.51, respectively. A study by Fassio et al. [83] of samples of ensiled maize kernels found similar values of R^2 for DM, CP, and ADF; however, the R^2 values for DMO and pH were higher (0.84 and 0.90, respectively). These authors also obtained an R^2 of 0.90 for the prediction of the content of ammonia nitrogen (NH_3-N). In this work, the use of the jack-knifing method improved the calibration models obtained. It is used to evaluate the stability of the calibrations and to eliminate nonsignificant wavebands in the calibration.

For biological parameters such as the *in vitro* digestibility (IVD) and metabolizable energy (ME) in pastures, Lobos et al. [84] established prediction equations with RMSEP values of 3.06 and 0.06 and R^2_p values of 0.90 and 0.94, respectively. The reliability of prediction of these NIRS parameters may be affected by the particle size, the effect of drying the sample prior to analysis, and the residual moisture in the samples after drying. Lovett et al. [85] studied such effects for samples of maize silage, with regard to the prediction of these parameters, concluding that the particle size was the most important factor, followed by the drying process and finally the presence of residual moisture. These authors used three statistical treatments—PLS, MPLS, and PCR—with two standard data preprocessing methods: standard normal variate (SNV) followed by detrending and first derivative. These authors used three statistical treatments—PLS, MPLS, and PCR—with two standard data preprocessing methods: standard normal variate (SNV) followed by detrending and first derivative.

Other authors have developed equations using NIRS to determine quality parameters that indicate whether the silage fermentation has been correct, and for quantification of fatty acids in forages. Thus, Sorensen et al. [86] developed prediction models (using PLS on full scan mean spectra after scatter correction with the standard normal variate (SNV) transformation combined with detrend and applying a second derivative) for the determination of lactic acid (Lac), acetic acid (HAc), pH, and NH3-N in maize silage. The RMSECV values were 4.7, 1.9, 2.4, and 2.9, respectively, and 4.0 for ethanol (EtOH). These authors showed that the NIRS technique is less accurate for HAc, but provides an estimate of its concentration. With regard to the quantification of fatty acids in fodders, Foster et al. [87] obtained high coefficients of determination for calibration (0.93–0.99) and cross-validation (0.89–0.98). The SEC and SECV were 20% lower compared to the mean. The RPD_{CV} was greater than 3 for all fatty acids except C12:0 (2.6) and C14:0 (2.9). The reliability of the prediction was lower, but acceptable for C12:0, C14:0, C18:0, C16:1, and C18:1. In this study, two limits were used for the validation of the prediction equations: GH (global spectral distance) and NH (neighborhood spectral distance) to determine if significant bias occurs and if there is a significant increase in unexplained error.

Finally, undesirable substances of a toxic nature can be found in animal forages. Fox et al. [88] established calibration equations for the estimation of hydrogen cyanide in forage sorghum. The equations developed by MLR gave a coefficient of determination (R^2) of 0.847 and an SEC value of 0.050%, with R^2 and SEP values for the validation of 0.829 and 0.057%, respectively. These authors found two important wavelengths for the prediction: 2034 and 2458 nm, associated with the former C=O carbonyl stretch and the latter associated with C-N-C stretching.

4.6. Organic residues and compost

The addition of value to wastes that are organic in nature is required to help reduce the increasing pollution, to optimize the use of available resources, and to offset the increasing energetic and economic costs of synthetic fertilizers. For this it is essential to know in detail the nature and type of such wastes, which can be used both fresh and stabilized. In this section we focus on fresh organic waste and stabilized materials—compost—resulting from the composting process. The origin of these organic wastes can be varied, but generally they originate from urban solid waste, sewage sludge generated in waste water treatment plants, and the agro-industrial sector.

The agri-food industry is one of the most important sectors in Europe and therefore the wastes it generates pose a serious environmental problem. Most of these organic wastes are considered biodegradable and derive from plants and animals. They include fruit and vegetable remains and crop pruning waste. Since these materials are organic or have a high organic matter content, usually with significant amounts of macronutrients as well, it is convenient to use them as organo-mineral fertilizers. This represents a double energy saving: first, waste is eliminated and, second, the need for synthetic fertilizers, whose cost has increased in recent years, is diminished [88].

In the treatment of organic wastes their possible uses must be taken into account, as the treatment determines the characteristics of the final product obtained. The fate of these organic wastes has been and is still very varied, depending on geographical location, activities taking place in a region, the population, facilities for reuse, and current regulations governing their handling and use. They can be used as soil conditioners, allowing long-term improvement of the physical properties of soils, reducing erosion, and helping the recovery of unproductive marginal areas. Another possibility is their use as substrates for the production of ornamental and horticultural plants. This requires improvement of the physical characteristics of the sludge or residue in question, which is achieved by composting [89].

Fresh organic wastes can be recycled by composting, a controlled bio-oxidative process involving numerous and varied microorganisms and requiring adequate moisture and heterogeneous organic substrates in the solid state. It involves a thermophilic stage and a temporary production of phytotoxins, giving—as the end-products of the degradation processes—carbon dioxide, water, minerals, and stabilized organic matter, free of phytotoxins and ready for use in agriculture without the risk of adverse phenomena. Finally, the compost, that can be defined as the product resulting from the composting process and maturation and that consists of stabilized organic matter like humus, is obtained. It has little resemblance to the original organic material as it will have been degraded, resulting in finer and dark particles. It is a product that is safe and free of phytotoxic substances, whose application to the soil will not cause damage to plants and which can be stored without further treatment or alterations [90, 91].

Thus, NIRS is used to predict different parameters and/or mineral elements in different organic residues and compost. The NIRS calibration results used successfully by different authors working with different types of organic matrices (industrial compost, compost of various animal manures, compost based on sludge and vegetable waste, compost based on winery and agro-industrial waste, compost derived from tofu waste and sewage sludge) show the great interest and the extent of use of this technique in the study of different variables in this type of organic material. In this regard there are several studies of compost or organic waste which highlight that the information generated with NIRS can increase the effectiveness of composting as a management method, due to the advantages that this technique presents as we have already discussed throughout this chapter. Thus, NIRs has been used to determine the next parameters.

4.6.1. Contents of mineral nutrients in compost

Mineral analysis of the materials at the start, during, and at the end in the composting process is desirable to ensure that the input materials are within acceptable ranges. To determine the mineral nutrient content it is necessary to make a mineral analysis of the compost material by ICP-OES and AA. In addition, nitrogen is one of the most important nutrients in the compost. When we analyze its total content, we refer to the sum of inorganic forms (ammonium, nitrate, and nitrite; NH_4^+, NO_3^-, NO_2^-, respectively) and organic (amino acids, proteins, nucleic acids, and other organic compounds having nitrogen in their structure). The usual methods for its analysis have been the Kjeldahl method (wet digestion) and the Dumas method (dry digestion). Although these conventional methods have been optimized, the digestion of the sample is still carried out with sulfuric acid and a series of catalysts which causes problems such as the emission of acid gases into the environment. Currently, near-infrared reflectance spectrometry technique (NIR) is available today and can be used to determine mineral nutrient content in organic residues and compost [91, 92]. Malley et al. [91] used a field-portable Corona 45 VIS NIR (visible/near-infrared) spectrometer (Carl Zeiss, Germany) from 360 to 1690 nm to measure total N, ammonium-N, organic N, P, K, Ca, Mg, S, Mn, Zn, and Cu in manure coming from beef cattle manure, stockpiled manure, and compost. The calibrations were developed for each constituent separately by using PCA/PLS1 in The Unscrambler. The calibrations were successfully developed for all parameter measurements (except available P, nitrate + nitrite, or Na). Therefore, field-portable NIRS offers a considerable advance over existing field and laboratory methods by providing rapid, comprehensive compositional analysis when and where the information is required to assist with management of the nutrients of cattle manure.

Usually, heavy metals are also measured in the composts as they can cause toxicities in the plants (Cu, Hg, Cd, Ni). In several publications, it has been observed that NIR technique can be successful to measure these metals in compost coming from raw material rich in these metals [88, 93]. An interesting study was carried out by Shen et al. [94], who investigated the use of NIR to detect copper (Cu) in animal manure. A total of 118 pig manure samples were collected from four provinces in China, and spectra were acquired in the range of 10,000–4000 cm^{-1}. Results showed that the prediction of Cu concentration in pig manure was feasible (r^2 = 0.84, RMSE = 198 mg/kg; SE/SD = 2.4). Although the heavy metals in the vis-NIR region do not absorb energy, Cu in pig manure can be detected by NIR spectroscopy because a high percentage of the Cu is complexed with $CONH_2$ or CONHR functional groups of organic ligands such as protein, urea, amino acids, and other amide compounds.

4.6.2. Organic matter and total organic carbon

Compost maturity has often been associated with the degree of compost humification. Compost stability refers to the degree to which composts have been decomposed to more stable organic materials. Various global parameters have been currently used to assess both maturation process and quality of the final product, including physicochemical properties, such as C:N ratio, humified organic and water-soluble carbon, and cation exchange capacity. Methods for measuring total C in soils, such as wet combustion or dry combustion, are generally very accurate, but too slow or costly for everyday analysis. It has been measured by NIR the contents of carbon and nitrogen in sewage sludge and green waste compost [95], and sewage sludge [96] with successful results. So, Albrecht et al. [95] analyzed changes in composts of sewage sludges and green wastes by NIRS technique of six stages of composting: 8, 20, 35, 75, 135, and 180 days. Maturity of compost was assessed through changes in

C:N ratio. Results of spectroscopic properties (200 wavelengths) were studied with several multivariate analyses showing a precise calibration models between spectral data, the C, N, C:N values, and composting time were build using partial least square regression ($r^2 > 0.95$). Together, these results show the efficiency of NIRS to predict chemical changes and the stage of transformation of organic matter during the composting process.

Humic acids from sewage sludge. Humic acids are part of the stable organic matter fraction in soils and composts. Due to their favorable properties for soils and plants, and their role in carbon sequestration, they are considered a quality criterion of composts. The traditional methodology for determining the content of humic acids is based on the solubility of the humic substances in aqueous media of different pH, i.e., humins are insoluble in any pH range, humic acids are insoluble in acid medium, and fulvic acids are soluble throughout the pH range. In the determination of humic acids by NIR in compost, a correlation coefficient of 0.94 and a standard error of estimation of 0.28 were obtained, values that can be considered very acceptable [97]. Other publications in mushroom compost [98], manure [99], and fattening pig manure [100] have given excellent results to characterize the humic acids in this material.

In summary, numerous studies of compost or organic wastes using NIRS have demonstrated the efficacy of this methodology. For all the above reasons, this spectroscopic tool is an emerging technique in the analysis of environmental parameters. It offers several advantages over traditional analytical techniques, such as rapidity, ease of preparation and handling of samples (no reagents are required), and low cost.

Nomenclature

ANN	Artificial Neural Networks
HAc	Acetic acid
ADF	Acid detergent fiber
ADIN	Acid detergent insoluble nitrogen
ADFCP	Acid detergent insoluble protein
CEC	Cation exchange capacity
R^2/r^2	Coefficient of determination
$R^2c/R^2p/R^2cv$	Coefficient of determination of the calibration/prediction/cross-validation
CP	Crude protein
Cys	Cystine
DT	Detrending

DM	Dry matter
DDM	Digestible dry matter
DNDF	Digestible neutral detergent fiber
DOM	Digestible organic matter
EtOH	Ethanol
FD	First derivate
GH	Global spectral distance
INDF	Indigestible neutral detergent fiber
IVD	In vitro digestibility
IAD	Index of Absorbance Difference
DRIS	Integrated System of Diagnosis and Recommendation
Lac	Lactic acid
LOS	Linear offset subtraction
Lys	Lysine
H	Mahalanobis distance values
MMN	Maximum-minimum normalization
ME	Metabolizable energy
Met	Methionine
MPLSR/MPLS	Modified partial least-squares regression
MLR	Multiple linear regression
MSC	Multiplicative scatter correction
MARS	Multivariate adaptive regression splines
NIRS	Near infrared reflectance spectroscopy
NH	Neighborhood spectral distance
NDF	Neutral detergent fiber
NDFCP	Neutral detergent insoluble protein
NDSC	Neutral detergent-soluble carbohydrates
NDSF	Neutral detergent-soluble fiber
NDP	No data preprocessing
NPN	Nonprotein nitrogen

NMSI	Normalized soil moisture index
PLSR/PLS	Partial least squares regression
PCR	Principal components regression
A	Relative absorbance values
RPD	Residual prediction deviation
RPDc/RPDp/RPDcv	Residual prediction deviation of the calibration/prediction/ cross-validation
RMSECV	Root mean square error of cross validation
RMSEE	Root mean square error of estimation
RMSEP	Root mean square error of prediction
SED	Second derivate
SWIR	Shortwave infrared
SN	Soluble nitrogen
TSS	Soluble solids content
SOM	Soil organic matter
SS	Spectral smoothing
SEC	Standard error of calibration
SECV	Standard error of cross-validation
SEP	Standard error of prediction
SNV/SNV-D	Standard normal variate, detrending
SR	Sufficiency range
SSE	Squared sum estimation
t-test	Student's *t*-test
SSL	Subtraction of a straight line
Thr	Threonine
TA	Titratable acidity
TN	Total nitrogen
TCP	True protein
Trp	Tryptophan
VN	Vector normalization

Author details

Francisco García-Sánchez[1]*, Luis Galvez-Sola[2], Juan J. Martínez-Nicolás[3], Raquel Muelas-Domingo[2] and Manuel Nieves[2]

*Address all correspondence to: fgs@cebas.csic.es

1 Department of Plant Nutrition, CEBAS-CSIC, Murcia, Spain

2 Department of Agrochemistry and Environment, EPSO-UMH, Orihuela, Spain

3 Department of Plant Production and Microbiology, EPSO-UMH, Orihuela, Spain

References

[1] Burns DA, Ciurczak EW. Handbook of near-infrared analysis. Volume 13 in Practical Spectroscopy Series. Marcel Dekker. Inc., New York. 1992.

[2] Gislum R, Micklander E, Nielsen JP. Quantification of nitrogen concentration in perennial ryegrass and red fescue using near-infrared reflectance spectroscopy (NIRS) and chemometrics. Field Crops Res. 2004;**88**:269–277.

[3] Bertrand D, Dufour E. Infrared Spectroscopy and its analytical applications. Editions TEC & DOC, Paris, France. 2000.

[4] Iwamoto M, Application of near-infrared spectroscopy for quality control in food and feed industries. In Proceedings of the 3rd International Conference on Near Infrared Spectroscopy. Biston R and Bartiaux-Thill N (eds.). Agricultural Research Centre, Gembloux, Belgium. 1991; 175–186.

[5] Shenk JS, Workman JJ Jr, Westerhaus MO. Application of NIR spectroscopy to agricultural products. in: D.A. Burns, E.W. Ciurczak (Eds.) Handbook of Near-Infrared Analysis. Marcel Dekker, Inc, New York; 1992; 383–431

[6] Barnes RJ, Dhanoa MS, Lister SJ. Standard normal variate transformation and de-trending of near-infrared diffuse reflectance spectra. Appl. Spectrosc. 1989;**43**:772–777.

[7] Dhanoa MS, Lister SJ, Sanderson R, Barnes RJ. The link between Multiplicative Scatter Correction (MSC) and Standard Normal Variate (SNV) transformations of NIR spectra. J. Near Infrared Spectrosc. 1994;**2**:43–47.

[8] Fernández-Cabanás V, Garrido-Varo A. The use of "standard normal variate" and "detrending" as signal improvement tools for the interpretation of near infrared spectra of agro-foodproducts. Química Analítica. 1999;**18**:113–118.

[9] Naes T, Issaksson T, Fearn T, Davies T. A user-friendly guide tomultivariatecalibration and classification. NIR Publications, Chichester, UK. 2002.

[10] Savitzky A, Golay MJE. Smoothing and differentiation of data by simplified least squares procedures. Anal. Chem. 1964;**36**:1627–1639.

[11] Norris KH, Williams PC. Optimization of mathematical treatments of raw near-infrared signal in the measurement of protein in hard red spring wheat. I. Influence of particle-size. Cereal Chemistry 61. 1984;**2**:158–165.

[12] Bertrand D. Data pre-treatment and original analysis in spectroscopy. Advanced Comet Chemometrics School, Libramont, Belgium. 1993; 26–28.

[13] Horlick G. Digital data handling of spectra utilizing Fourier transformations. Anal. Chem. 1972;**44**:943–947.

[14] Siebielec G, McCarty GW, Stuczynski TI, Reeves JB. Near- and mid-infrared diffuse reflectance spectroscopy for measuring soil metal content. J. Environ. Qual. 2004;**33**:2056–2069.

[15] Shenk JS, Westerhaus MO. Analysis and Food Products by Near Infrared Reflectance Spectroscopy. Monograph. NIRSystems. 1995.

[16] Martens H, Naes T. Multivariate calibration. John Wiley and Sons Inc., Chichester, UK. 1989.

[17] Wold S, Sjöström M, Eriksson L. PLS-regression: a basic tool of chemometrics. Chemometr. Intell. Lab. Syst. 2001;**58**:109–130.

[18] Massart DL, Vandeginste BGM, Buydens LMC. Chemometrics: A Textbook. Vol. 2. Ed. Elsevier, Amsterdam. 1988.

[19] Stone M. Cross validation choice and assessment of statistical prediction. J. R. Statist. Soc. B. 1974;**39**:111–133

[20] Williams PC, Sobering D. How do we do it: a brief summary of the methods we use in developing near infrared calibrations. In: Near Infrared Spectroscopy: The Futures Waves. Davies AMC and Williams PC (eds.). NIR Publications, Chischester, UK. 1996, 185–188.

[21] Shenk JS, Westerhaus MO. Calibration the ISI way. In: Davies AMC and Williams P (eds.). NIR Publications, Chichester, UK. 1996, 198–202.

[22] Shenk JS, Workman JJ Jr, Westerhaus MO. Application of NIR spectroscopy to agricultural products. in: D.A. Burns, E.W. Ciurczak (Eds.) Handbook of Near-Infrared Analysis. Marcel Dekker, Inc, New York; 1992; 383–431.

[23] Fearn T. The chemometric space. NIR News. 1987.

[24] Malley DF, Ben-Dor E, Martin PD. Application in analysis of soils. In: Robert C, Workman J, Reeves JB (eds). Near Infrared Spectroscopy in Agriculture. American Society of Agronomy, Soil Science Society of America and Crop Science Society of America, Madison, WI. 2004, 729–784.

[25] Williams PC. Implementation of Near-infrared Technology. In: Near-infrared technology in the agricultural and food industries. Williams P, and Norris K (eds.) (2nd ed.). American Association of Cereal Chemists, St. Paul, MN, USA. 2001, 296.

[26] Windham WR, Mertens DR, Barton FE. II Protocol for NIRS calibration: Sample selection and equation development and validation. In: Marten CG, Shenk JS, and Barton FE

(eds.), Near Infrared Reflectance Spectroscopy (NIRS): Analysis of Forage Quality (643rd ed.). USDA-ARS: Agricultural Handbook, Washington, DC. 1989, 96–103.

[27] Walsh LM. Soil and Applied Nitrogen. Soil and Applied Phosphorus. Soil and Applied Potassium. Soil and Applied Boron. Soil and Applied Calcium. Soil and Applied Magnesium. University of Wisconsin Extension Bulletin A2519 to A2524. University Wisconsin Extension, Madison, WI. 1973.

[28] Huang G, Han L, Yang Z, Wang X. Evaluation of the nutrient metal content in Chinese animal manure compost using near infrared spectroscopy (NIRS). Biores. Technol. 2008;**99**:8164–8169.

[29] Garnsworthy PC, Wiseman J, Fegeros K. Prediction of chemical, nutritive, and agronomic characteristics of wheat by near infrared spectroscopy. J. Agric. Sci. 2000;**135**:409–417.

[30] Zhang GC, Li Z, Yan XM, Cheng CG, Zhou P, Lin GL, Zhou GJ, Liu N, Han XR. Rapid analysis of apple leaf nitrogen using near infrared spectroscopy and multiple linear regression. Commun. Soil Sci. Plant Anal. 2012;**43**:1768–1772.

[31] Halgerson JL, Sheaffer CC, Martin NP, Peterson PR, Weston SJ. Near-infrared reflectance spectroscopy prediction of leaf and mineral concentrations in alfalfa. Agron. J. 2004;**96**:344–351.

[32] Yarce CY, Rojas G. Near infrared spectroscopy for the analysis of macro and micro nutrients in sugarcane leaves. Sugar Industry. 2012;**11**:707–710.

[33] Lebot V, Malapa R, Jung M. Use of NIRS for the rapid prediction of total N, minerals, sugars and starch in tropical root and tuber crops. New Zeal. J. Crop Hort. Sci. 2013;**41**:144–153.

[34] Rossa ÜB, Angelo AC, Nisgoski S, Westphalen DJ, Frizon CNT, Hoffmann-Ribani R. Application of the NIR method to determine nutrients in yerba mate (*ilex paraguariensis* a. st.-hill) leaves. Commun. Soil Sci. Plant Anal. 2015;**46**:2323–2331.

[35] Min M, Lee WS, Burks TF, Jordan JD, Schumann AW, Schueller JK, Xie H. Design of a hyperspectral nitrogen sensing system for orange leaves. Comput. Electron. Agric. 2008;**63**:215–226.

[36] Menesatti P, Antonucci F, Pallotino F, Roccuzzo G, Allegra M, Stagno F, Intringliolo F. Estimation of plant nutritional status by Vis-NIR spectrophotometric analysis on orange leaves [*Citrus sinensis* (L) Osbeck*cv*Tarocco]. Biosyst. Eng. 2010;**105**:448–454.

[37] Galvez-Sola L, Garcia-Sanchez F, Perez-Perez JG, Gimeno V, Navarro JM, Moral R, Martinez-Nicolas JJ, Nieves M. Rapid estimation of nutritional elements on citrus leaves by near infrared reflectance spectroscopy. Front. Plant Sci. 2015;**6**:571.

[38] Xuemei L, Jianshe L. Using short wave visible-near infrared reflectance spectroscopy to predict soil properties and content. Spectros. Lett. 2014;**47**:729–739. DOI: 10.1080/00387010.2013.840315

[39] Islam K, Singh B, McBratney A. Simultaneous estimation of several soil properties by ultra-violet, visible, and near-infrared reflectance spectroscopy. Aust. J. Soil Res. 2003;**41**:1101–1114.

[40] Stenberg B, Viscarra Rossel RA, Mouazen AM, Wetterlind J. Visible and near infrared spectroscopy in soil science. In Sparks DL (ed.), Advances in Agronomy, Vol. 107. Academic Press, Burlington. 2010, 163–215. http://dx.doi.org/10.1016/S0065-2113(10)07005-7

[41] Kuang BY, Mouazen AM. Non-biased prediction of soil organic carbon and total nitrogen with vis-NIR spectroscopy, as affected by soil moisture content and texture. Biosyst. Eng. 2013,**114**:249–258.

[42] Wight JP, Ashworth AJ, Allen FL. Organic substrate, clay type, texture, and water influence on NIR carbon measurements. Geoderma. 2016;**261**:36–43.

[43] Udelhoven T, Emmerling C, Jarmer T. Quantitative analysis of soil chemical properties with diffuse reflectance spectrometry and partial least-square regression: A feasibility study. Plant Soil. 2003:**251**:319–329.

[44] Martin PD, Malley DF, Manning G, Fuller L. Determination of soil organic carbon and nitrogen at the field level using near-infrared spectroscopy. Can. J. Soil Sci. **2002**;82:413–422.

[45] Buddenbaum H, Steffens M. The effects of spectral pretreatments on chemometric analyses of soil profiles using laboratory imaging spectroscopy. Appl. Environ. Soil Sci. 2012;2012:1–12. doi: 10.1155/2012/274903.

[46] Bilgili AV, Cullu MA, Es HV, Aydemir A, Aydemir S. The use of hyperspectral visible and near infrared reflectance spectroscopy for the characterization of salt-affected soils in the Harran plain, Turkey. Arid Land Res. Manage. 2011,**25**:19–37.

[47] Mashimbye ZE, Cho MA, Nell JP, De Clercq WP, Van Niekerk A, Turner DP. Model-based integrated methods for quantitative estimation of soil salinity from hyperspectral remote sensing data: A case study of selected South African soils. Pedosphere. 2012;**22**:640–649.

[48] Shi Z, Ji W, Rossel RAV, Chen S, Zhou Y. Prediction of soil organic matter using a spatially constrained local partial least squares regression and the Chinese vis-NIR spectral library. Eur. J. Soil Sci. 2015;**66**:679–687.

[49] Stenberg B, Jonsson A, Borjesson T. Near infrared technology for soil analysis with implications for precision agriculture. In: Davies A and Cho R (eds.), Near Infrared Spectroscopy: Proceedings of the 10th International Conference. NIR Publications, Chichester, UK. 2002, 279–284.

[50] Wetterlind J, Stenberg B, Jonsson A. Near infrared reflectance spectroscopy compared with soil clay and organic matter content for estimating within-field variation in N uptake in cereals. Plant Soil. 2008;**302**:317–327.

[51] Nawar S, Buddenbaum H, Hill J, Kozak J, Mouazen AM. Estimating the soil clay content and organic matter by means of different calibration methods of vis-NIR diffuse reflectance spectroscopy. Soil Tillage Res. 2016;**55**:510–522.

[52] Nocita M, Stevens A, Noon C, van Wesemael B. Prediction of soil organic carbon for different levels of soil moisture using Vis-NIR spectroscopy. Geoderma. 2013;**199**:37–42.

[53] Viscarra Rossel RA, McGlynn RN, McBratney AB. Determining the composition of mineral-organic mixes using UV-vis-NIR diffuse reflectance spectroscopy. Geoderma. 2006;**137:**70–82.

[54] Viscarra Rossel RA, Walvoort DJJ, McBratney AB, Janik LJ, Skjemstad JO. Visible, near infrared, mid infrared or combined diffuse reflectance spectroscopy for simultaneous assessment of various soil properties. Geoderma. 2006;**131**:59–75.

[55] Viscarra Rossel RA, Cattle SR, Ortega A, Fouad Y. *In situ* measurements of soil colour, mineral composition and clay content byvis-NIR spectroscopy. Geoderma. 2009;**150**:253–266.

[56] Wu CY, Jacobson AR, Laba M, Kim B, Baveye PC. Surrogate correlations and near-infrared diffuse reflectance sensing of trace metal content in soils. Water Air Soil Pollut. 2010;**209**:377–390.

[57] Vendrame PRS, Marchao RL, Brunet D, Becquer T. The potential of NIR spectroscopy to predict soil texture and mineralogy in Cerrado Latosols. Eur. J. Soil Sci. 2012;**63**:743–753.

[58] Ben-Dor E, Banin A. Near infrared analysis (NIRA) as a method to simultaneously evaluate spectral featureless constituents in soils. Soil Sci. 1995;**159**:259–270.

[59] Curcio D, Ciraolo G, D'Asaro F, Minacapilli M, Romano N, DUrso G, Severino G, Chirico GB, Palladino M. Prediction of soil texture distributions using VNIR-SWIR reflectance spectroscopy. Four decades of progress in monitoring and modeling of processes in the soil-plant-atmosphere system: applications and challenges. Procedia Environ. Sci. 2013;**19**:494–503.

[60] Chapman HD. Cation exchange capacity. In: Black CA (eds.), Methods of Soil Analysis, Part 2, Number 9 in the Series. Agronomy: Am. Inst. Agronomy, Madison, Wisconsin. 1965, 891–901.

[61] Ulusoy Y, Tekin Y, Tumsavas Z, Mouazen AB. Prediction of soil cation exchange capacity using visible and near infrared spectroscopy. Biosyst. Eng. 2016;**XX**:1–15.

[62] Chang CW, Laird DA, Mausbach MJ, Hurburgh CR. Near-infrared reflectance spectroscopy-principal components regression analyses of soil properties. Soil Sci. Soc. Am. J. 2001;**65**:480–490.

[63] Todorova M, Mouazen AM, Lange H, Atanassova S. Potential of near-infrared spectroscopy for measurement of heavy metals in soil as affected by calibration set size. Water Air Soil Pollut. 2014;**225**:8.

[64] Okparanma RN, Coulon F, Mayr T, Mouazen AM. Mapping polycyclic aromatic hydrocarbon and total toxicity equivalent soil concentrations by visible and near-infrared spectroscopy. Environ. Pollut. 2014;**192**:162–170.

[65] Bullock PR, Li X, Leonardi L. Near-infrared spectroscopy for soil water determination in small soil volumes. Can. J. Soil Sci. 2004;**84**:333–338.

[66] Chang GW, Laird DA, Hurburgh GR. Influence of soil moisture on near-infrared reflectance spectroscopic measurement of soil properties. Soil Sci. 2005;**170**:244–255.

[67] Pu YY, Feng YZ, Sun DW. Recent progress of hyperspectral imaging on quality and safety inspection of fruits and vegetables: a review. Comp. Rev. Food Sci. Food Safety. 2015;**14**:176–188.

[68] Sun XD, Liu YD, Li YF, Wu MM, Zhu DN. Simultaneous measurement of brown core and soluble solids content in pear by on-line visible and near infrared spectroscopy. Postharvest Biol. Technol. 2016;**116**:80–87.

[69] Ribera-Fonsecaa A, Noferini M, Jorquera-Fontenad E, Rombolà AD. Assessment of technological maturity parameters and anthocyanins in berries of cv. Sangiovese (*Vitis vinifera* L.) by a portable vis/NIR device. Scientia Horticulturae. 2016;**209**:229–235.

[70] Fernandez-Espinosa AJ. Combining PLS regression with portable NIR spectroscopy to on-line monitor quality parameters in intact olives for determining optimal harvesting time. Talanta. 2016;**148**:216–228.

[71] Rubenthaler GL, Bruinsman BL. Lysine estimation in cereals by NIR. Crop Sci. 1978;**18**:1039–1042.

[72] Fontaine J, Schirmer B, Horr, J. Near- infrared reflectance spectroscopy (nirs) enables the fast and accurate prediction of essential amino acid contents. results for wheat, barley, corn, triticale, wheat bran/middlings, rice bran, and sorghum. J. Agric. Food Chem. 2002;**50**:3902–3911.

[73] Kovalenko IV, Rippke GR, Hurburgh CR. Determination of amino acid composition of soybeans (Glycine max) by near-infrared spectroscopy. J. Agric. Food Chem. 2006;**54**:3485–3491.

[74] Ferreira DS, Pallone JAL, Poppi RJ. Fourier transform near-infrared spectroscopy (FT-NIRS) application to estimate Brazilian soybean [Glycine max (L.) Merril] composition. Food Res. Int. 2013;**51**:53–58.

[75] Wang J, Liu H, Ren G. Near-infrared spectroscopy (NIRS) evaluation and regional analysis of Chinese faba bean (*Vicia faba* L.). Crop J. 2014;**2**:28–37.

[76] Sánchez-Vioque R, Rodríguez-Conde MF, De los Mozos Pascual M. Application of near-infrared reflectance spectroscopy for the estimation of protein and L-canavanine contents in seeds of one-flowered vetch (Vicia articulata Hornem). Span. J. Agric. Res. 2009;**7**:645–651.

[77] Berardo N, Pisacane V, Battilani P, Scandolara A, Pietri A, Marocco A. rapid detection of kernel rots and mycotoxins in maize by near-infrared reflectance spectroscopy. J. Agric. Food Chem. 2005;**53**:8128–8134.

[78] Volkers KC, Wachendorf M, Loges R, Jovanovic NJ, Taube F. Prediction of the quality of forage maize by near-infrared reflectance spectroscopy. Anim. Feed Sci. Technol. 2003;**109**:183–194.

[79] Hermida M, Lois A, Rodriguez-Otero J.L. Analysis of nitrogen fractions in silage by near-infrared spectroscopy. J. Agric. Food Chem. 2005;**53**:1374–1378.

[80] Nie Z, Han J, Liu T, Liu X. Hot topic: Application of support vector machine method in prediction of alfalfa protein fractions by near infrared reflectance spectroscopy. J. Dairy Sci. 2008;**91**:2361–2369.

[81] Nousiainen J, Ahvenjärvi S, Rinne M, Hellämäki M, Huhtanen P. Prediction of indigestible cell wall fraction of grass silage by near infrared reflectance spectroscopy. Anim. Feed Sci. Technol. 2004;**115**:295–311.

[82] Cozzolino D, Fassio A, Fernández E, Restaino E, La Manna A. Measurement of chemical composition in wet whole maize silage by visible and near infrared reflectance spectroscopy. Anim. Feed Sci. Technol. 2006;**129**:329–336.

[83] Fassio A, Fernández EG, Restaino EA, La Manna A, Cozzolino D. Predicting the nutritive value of high moisture grain corn by near infrared reflectance spectroscopy. Comput. Electron. Agric. 2009;**67**:59–63.

[84] Lobos I, Gou P, Hube S, Saldaña R, Alfaro M. Evaluation of potential of nirs to predict pastures nutritive value. J. Soil Sci. Plant Nutr. 2013;**13**:463–468.

[85] Lovett DK, Deaville ER, Givens DI, Finlay M, Owen E. Near infrared reflectance spectroscopy (NIRS) to predict biological parameters of maize silage: effects of particle comminution, oven drying temperature and the presence of residual moisture. Anim. Feed Sci. Technol. 2005;**120**:323–332.

[86] Sorensen LK, et al. Prediction of fermentation parameters in grass and corn silage by near infrared spectroscopy. J. Dairy Sci. 2004;**87**:3826–3835.

[87] Foster JG, Clapham WM, Fedders JM. Quantification of fatty acids in forages by near-infrared reflectance spectroscopy. J. Agric. Food Chem. 2006;**54**:3186–3192.

[88] Fox GP, O'Donnell NH, Stewart PN, Gleadow RM. estimating hydrogen cyanide in forage sorghum (Sorghum bicolor) by near-infrared spectroscopy. J. Agric. Food Chem. 2012;**60**:6183–6187.

[89] Perez-Murcia MD, Moral R, Moreno-Caselles J, Perez-Espinosa A, Paredes C. Use of composted sewage sludge in growth media for broccoli. Bioresour. Technol. 2006;**97**:123–130.

[90] Costa F, García C, Hernández T, Polo A. Residuos Orgánicos Urbanos. Manejo y Utilización. Ed. CSIC-CEBAS, Murcia. 1991.

[91] Malley DF, McClure C, Martin PD, Buckley K, McCaughey WP. Compositional analysis of cattle manure during composting using a field portable near infrared spectrometer. Commun. Soil Sci. Plant Anal. 2005;**36**:455–475.

[92] Fujiwara T, Murakami K. Application of near-infrared reflectance spectroscopy for estimating available nitrogen in poultry manure compost. Soil Sci. Plant Nutr. 2007;**53**:102–107.

[93] Moral R, Galvez-Sola L, Moreno-Caselles J, Perez-Murcia MD, Perez-Espinosa A, Paredes C. Can near-infrared reflectance spectroscopy (NIRS) predict heavy metals in sewage sludge? In: Kungolos A, Aravossis K, Karagiannidis A, and Samaras P (eds.), First Conference on Environmental Management, Engineering, Planning and Economics. Skiathos Island, Greece. 24–28 June 2007, 1683–1688.

[94] Shen X, Yang Z, Huang G, Han L. Near infrared spectroscopy detection of copper in pig manure and the spectral basis of the analysis. J. Near Infrared Spectrosc. 2014;**22**:305–312.

[95] Albrecht R, Joffre R, Gros R, Le Petit J, Terrom G, Périssol C. Efficiency of near-infrared reflectance spectroscopy to assess and predict the stage of transformation of organic matter in the composting process. Bioresour. Technol. 2008;**99**:448–455.

[96] Vergnoux A, Guiliano M, Le Dréau Y, Kister J, Dupuy N, Doumenq. Monitoring of the evolution of an industrial compost and prediction of some compost properties by NIR spectroscopy. Sci. Total Environ. 2009;**407**:2390–2403.

[97] Polak J, Sulkowski WW, Bartoszek M, Papiez W. Spectroscopic studies of the progress of humification processes in humic acid extracted from sewage sludge. J. Mol. Struct. 2005;**744**:983–989.

[98] Sharma HSS, Kilpatrickb M, Burns L. Determination of phase II mushroom (*Agaricus bisporus*) compost quality by near infrared spectroscopy. J. Near Infrared Spectrosc. 2000;**8**:1–19.

[99] Xing L, Chen LJ, Han LJ. Rapid analysis of layer manure using near-infrared reflectance spectroscopy. Poult. Sci. 2008;**87**:1281–1286.

[100] Yang JC, Maie N, Tada Y, Katayama A. Characterization of the maturing process of cattle manure compost. Process Biochem. 2006;**41**:380–389.

Chapter 6

Near-Infrared Spectroscopy Combined with Multivariate Tools for Analysis of Trace Metals in Environmental Matrices

Philiswa N. Nomngongo,
Tshimangadzo S. Munonde, Anele Mpupa and
Nkositetile Raphael Biata

Additional information is available at the end of the chapter

http://dx.doi.org/10.5772/67199

Abstract

Environmental contamination by trace elements is becoming increasingly important problem worldwide. Trace metals such as cadmium, copper, lead, chromium, and mercury are major environmental pollutants that are predominantly found in areas with high anthropogenic activities. Therefore, there is a need for rapid and reliable tools to assess and monitor the concentration of heavy metal in environmental matrices. A nondestructive, cost-effective, and environmentally friendly procedure based on near-infrared reflectance spectroscopy (NIRS) and chemometric tools has been used as alternative technique for the simultaneous estimation of various heavy metal concentrations in environmental sample. The metal content is estimated by assigning the absorption features of metals associated with molecular vibrations of organic and inorganic functional groups in organic matter, silicates, carbonates, and water at 780–2500 nm in the near-infrared region. This chapter, reviewed the application of NIRS combined with chemometric tools such as multiple linear regression (MLR), principal component regression (PCR), and partial least squares (PLS) regression. The disadvantages and advantages of each chemometric tool are discussed briefly.

Keywords: near-infrared spectroscopy, principal component regression, partial least squares, multiple linear regression, trace metals

1. Introduction

Due to fast industrial development and growth that have happened in most areas of the world during the recent years, water and soil are getting a large amount of pollutants such as trace elements from different sources [1]. This can, however, lead to environmental contamination, thus affecting the ecosystem. Contamination refers to the condition of the land or water where any chemical substance or waste has been added at the above background level. The signifies of water, air, and land pollution include an adverse health or environmental impact [2–4] run-off, aerial deposition of chemicals used for agriculture or industry, materials stored or dumped on the site, and contaminants in imported fill, and building demolition can also result in contamination of the soil and water that are close to residential communities [5]. Contaminants such as trace metals may be introduced into drinking water via the aforementioned activities or leached from the soil into groundwater [6]. Additionally, trace metals occur naturally in the earth's crust [7]. For this reason, they can be present in soils at a background level. Trace metals persist for a long time in the environment because they are not degradable. In addition, they are translocated to different components, thus affecting the biota [2–4]. The persistence of trace metals can result in bioaccumulation and biomagnifications causing heavier exposure for some living organisms that are present in the environment [8].

Trace metal contaminations threaten agriculture and other food sources for human population as well as poor vegetation growth and that lower plant resistance against pests [9]. This situation poses different kinds of challenges for remediation. Furthermore, people can be exposed to contaminants in the soil through different ways. These include dermal exposure or inhalation and penetration via the skin or eyes (includes exposure to dust) [5]. Trace metal exposure is normally chronic (exposure over a longer period of time), due to food chain transfer [2]. But the case of acute (immediate) poisoning is rare through ingestion or dermal contact but is possible [2]. The toxicity of trace metals is one of the major environmental health concerns and potentially dangerous due to bioaccumulation through the food chain [4].

In view of the abovementioned challenges, the development of sensitive and selective analytical procedure for the determination of trace metals is of great importance. Flame atomic absorption spectrometry (FAAS) [10], graphite furnace atomic absorption spectrometry (GFAAS) [11], cathodic and anodic stripping voltammetry [12, 13], inductively coupled plasma optical emission spectrometry (ICP-OES) [14], and inductively coupled plasma mass spectrometry (ICP-MS) [14] are the most widely used analytical techniques for determination of trace metals in different matrices. However, these techniques are expensive, tedious, complex, and highly time-consuming [15–17]. In addition, investigation of trace metal concentration distributions in environmental matrices is based on numerous samples and laboratory analysis. Therefore, a rapid, reliable, and environmentally friendly method is required to detect and survey the distribution of trace metals in environmental matrices. This is done in order to diagnose suspected contaminated areas as well as control the rehabilitation processes [18]. Reflectance spectroscopy is the study of the absorption and emission of light and other radiations by matter as related to the dependence of these processes on the wavelength of the radiation [19]. It is based on the distinct vibrations and electronic processes of chemical bonds in molecules [20]. These vibrations can be observed in three regions, namely, far IR (25×10^3 nm–1×10^6 nm), mid-IR (25×10^2 nm–25×10^3 nm), and near IR (8×10^2 nm–25×10^2 nm), with mid-IR and near IR being the most useful for qualitative and quantitative analysis [21]. The MIRS is known to have a greater predictive ability for soil geochemistry compared with the visible-near infrared (vis-NIR) [22, 23]. However, more interest is on NIRs because it is generally cheaper [23]. In NIR, many trace elements are spectrally featureless and only exhibit characteristic spectral features at high concentrations (>4000 mg kg^{-1}) [23, 24]. Therefore, due to high detection, low levels of heavy metals can be indirectly determined from spectra due to their association with Fe oxides, clays, and organic matter [24]. For this reason, cost-effective and nondestructive analytical techniques based on near-infrared (NIR) spectroscopy coupled with chemometrics have been developed to overcome the problems encounter when using traditional methods [17, 25].

The aim of this chapter is to review the application of near-infrared spectroscopy (NIRS) combined with chemometrics for the estimation of the concentration and distribution of trace elements in environmental matrices. The disadvantages of NIRS without chemometrics for analysis of trace elements are discussed. In addition, this chapter aims at promoting the application of simpler and greener methods such as the combination of NIRS and multivariate tools for monitoring of trace metal contaminations in different matrices.

2. Application of near-infrared spectroscopy (NIRS) for analysis of trace metals

Near-infrared spectroscopy (NIRS) is a fast and nondestructive analytical technique that is used to provide multi-constituent analysis of almost all types of sample matrices [15–17, 25]. This technique covers wavelength range closer to the mid-infrared and broadens up to the visible region [26, 27]. Typically, NIRS is primarily based on absorbance characteristics caused by vibrations of covalent bonds between H, C, O, S, and N, which are the main components of the organic matter [28]. Pure metals do not absorb in the NIR region [29]. However, their indirect detection is possible via their complexion with organic molecules containing C–H, N–H, and O–H bonds, which are detectable [30]. This concept was termed "aquaphotomics" which is based on fact that the characteristic absorbance pattern of water (O–H overtones) can change as a consequence of the binding reaction with the metal [31, 32]. This effect is described and demonstrated in the study carried out by Putra et al.; although some complexes might be similar in different samples, slight differences in spectral features such as shifts in peak wavelength may still be seen depending on the nature of the cation [30]. In addition, the electromagnetic radiation spectrum in the near-infrared region contains useful information about environmental sample constituents such as soil that can be used for prediction of metal concentrations [25, 28, 33]. For instance, the absorption features associated with electronic transitions of Fe^{3+} and Fe^{2+} ions in Fe-bearing minerals can be found in the near-infrared region at 780–1200 nm [33–35]. In addition, the absorption features of metals associated with molecular vibrations of organic and inorganic functional groups in organic matter, silicates, carbonates, and water can be found in near infrared at 780–2500 nm region [24, 33, 34, 36].

Wu et al. [24, 36] reported the feasibility of using NIRS for monitoring and predicting trace metals in suspended solids, sediment, and soil. This was achieved by quantitative evaluations of the spectral activity of sediment and soil properties [23–37]. However, due to challenges such as the collinearity, band overlaps, and interactions for some soil properties, the spectra of soil, sediment, or suspended solids are often broad and nonspecific [25]. To overcome these challenges, some chemometric tools have been used to be applied to the quantitative analysis of the spectroscopic data [38]. These chemometric tools include multiple linear regression (MLR) [39], principal component regression (PCR) [40, 41], and partial least squares (PLS) regression [42]. These chemometric tools have been used to characterize soil spectra and build models for estimating the trace metal concentrations in soil or sediments and other matrices [25].

2.1. Applications of NIRS-multi-linear regression determination trace metals in environmental samples

Multiple linear regression (MLR) is a conventional chemometric method that commonly correlates a linear combination of several selected spectral bands/indices which have high correlations with the heavy metal concentrations [18]. The disadvantage of using MLR is that it does not perform well with hyperspectral measurements. This is because the NIR spectral data usually exhibit high collinearity [18, 39]. This challenge has been solved by applying the enter and stepwise MLR approaches [18, 39]. In enter-MLR approach, a procedure for variable selection is adopted, and the selected variables are then used to calibrate the MLR model [18]. In stepwise-MLR approach, on the other hand, a forward or backward method is applied to progressively select the independent variables according to a tolerance significance level, which is generally set to 0.05 [18, 39]. Due to the challenges associated with MLS, there are very few reports on its application together with NIRS for determination of trace elements in environmental samples.

Kemper and Sommer [39] explored the possibility to adapt chemometrics approaches for the quantitative estimation of As, Cd, Cu, Fe, Hg, Pb, S, Sb, and Zn in polluted soils using stepwise multiple linear regression (MLR) analysis and an artificial neural network (ANN) approach. The authors reported that the models predicted six out of nine elements with high accuracy. In addition, it was discovered that most wavelengths important for prediction were attributed to absorption features of iron and iron oxides. Furthermore, their results revealed the feasibility to predict heavy metals in contaminated soils using the rapid and cost-effective NIRS. Other applications are reported by Malley et al. [40] and Choe et al. [41].

2.2. Applications of NIRS-PCR for determination trace metals in environmental samples

Principal component regression (PCR) is a chemometric tool that combines principal component analysis and MLR [18, 42]. In this method, the independent variables are first decomposed into orthogonal principal components using the nonlinear iterative partial least squares algorithm and full cross validation of the calibration set [18]. The maximum number of principal components is then defined according to the minimum value of the root-mean-square error of the cross validation [18]. In the final step, the chosen principal components are used to calibrate the MLR models [43]. The advantage of PCR over the normal MLR is that the principal components are uncorrelated and the noise is filtered [18]. There is very limited information on the application of NIRS-PCR on the analysis of trace metals in different matrices. However, some of the reports are available in the literature [42, 43].

Wu et al. [36] reported the practicality of using NIRS for the determination of Hg concentration agricultural soil samples. The accuracy of the prediction models was optimized by applying several spectral pretreatments to the reflectance spectra. The univariate regression and principal component regression were used for the prediction of Hg concentration. According to their results, the optimal model was achieved using the PCR combined with Kubelka-Munk transformation. In addition, the results obtained from the correlation analysis revealed that Hg concentration correlated negatively with soil reflectance, while positively with the absorption depths of goethite at 0.496 μm and clay minerals at 2.21 μm [36]. The findings suggest that the adsorption of Hg by clay-size mineral accumulations in soils was the mechanism that can be used to predict the spectral absorption band of Hg.

2.3. Applications of NIRS-partial least squares regression environmental samples

Partial least squares regression (PLSR) is a chemometric method that is widely used to quantitatively derive information from NIR spectra [18, 44]. The PLSR allows a refined statistical approach using the full spectral region rather than unique and isolated analytical bands [44]. The principle of PLSR is based on incorporation of the dependent variables in the calculation of the principal components [45, 46]. For this reason, the PLSR is able to handle data with strong collinearity and noise [18, 44]. In addition, PLSR provides the possibility of handling cases where the number of variables significantly exceeds the number of available samples [47]. The applications of NIRS combined PLSR for the determination of metals in different environmental matrices have been widely reported in the literature (see **Table 1**).

Moros et al. [15] evaluated the potential of near-infrared (NIR) diffuse reflectance infrared Fourier transform spectroscopy (DRIFTS) combined with PLSR for nondestructive determinations of trace elements in foods. This analysis was achieved without physical or chemical sample pretreatment. The authors compared two spectral pretreatments that are multiplicative signal correction (MSC) and standard normal variate (SNV). Their results revealed that the PLS models built after using SNV provided the best prediction results for the determination of arsenic and lead in powdered red paprika samples. The concluded results showed that NIR diffuse reflectance spectroscopy combined with the PLS could be used to estimate the concentration of As and Pb at 100 $\mu g\ kg^{-1}$ level with a standard error of prediction of 39 and 50 $\mu g\ kg^{-1}$ for As and Pb, respectively. Furthermore, the estimated percentage errors were lower than 25% without the need of using sophisticated and high-cost instrumentation (such As ICP-MS and GFAAS) together with tedious and expensive digestion procedures for sample preparation. Moreover, the suggested NIRS-PLSR methodology was found to be an important tool for screening of trace elements in foods in the laboratories [15].

In another study, Li et al. [48] reported a method for simultaneous determination of mercury, lead, and cadmium ions in water samples using solid-phase extraction and near-infrared diffuse reflectance spectroscopy (NIRDRS). In order to analyze trace metal content in water samples using NIRS, thiol-functionalized magnesium phyllosilicate (Mg-MTMS) was used as an adsorbent for extraction of target analytes from aqueous solution. The adsorbed metals were measured using NIRDRS combined with PLS models. This combination resulted in fast

Analytes	Matrix	Concentra ions	Ref.
Cr, Co, Ni, Cu, Zn, As, Se, Cd, and Tl	Soil	0.32–110 $\mu g\ g^{-1}$	[16]
Zn, Pb, and Cd	Soil	0.17–6530 $mg\ kg^{-1}$	[41]
Pb^{2+}, Zn^{2+}, Cu^{2+}, Cd^{2+}, and Cr^{3+}	Water		[42]
Cd and Zn	Soil	2.25–51.48 $mg\ L^{-1}$	[43]
K, Ca, Mg, Fe, and Zn	Manure compost	0.676–80.97 $mg\ kg^{-1}$	[44]
Cd, Cu, Zn, Pb, Ni, Mn, and Fe	Freshwater sediments	7.63–198.20 $g\ kg^{-1}$	[45]
Zn and Pb	Soil	2–425 $mg\ g^{-1}$	[46]
Al, Ag, As, Ba, Be, Bi, Ca, Cd, Ce, Cs, Co, Cr, Cu, Fe, Ga, In, K, La, Li, Mg, Mn, Mo, Na, Nb, Ni, P, Pb, Rb, S, Sb, Sc, Sn, Sr, Te, Th, Ti, Tl, U, V, W, Y, and Zn	Soil	1672–4601 $mg\ kg^{-1}$	[47]
Fe, Zn, Mn, and Cu	Agricultural soils	50–100 $mg\ kg^{-1}$	[48]
Cu, Mn, Zn, and Fe	Water and HNO_3	–	[49]
		1–10,000 $\mu g\ L^{-1}$	

Table 1. Applications of NIRS-PLS regression for determination of trace metals in environmental samples.

and simultaneous quantitative prediction. The metal ions interacted with a functional group of the adsorbent and their absorption bands were observed in the spectra, thus leading to an efficient and precise prediction models. The concentration of the three metals that can be correctly determined was found to range between 4 and 6 mg L^{-1}. This proved that the adsorbent used (thiol-functionalized magnesium phyllosilicate) had a high efficiency for the enrichment of Hg, Pb, and Cd in dilute solution. Furthermore, the results obtained revealed the feasibility of NIRDRS-PLSR for quantitative analysis metal ions in river water. Other applications of NIRS combined with PLSR are presented in **Table 1**.

2.4. Nonlinear calibration models for near-infrared spectroscopy

The abovementioned linear calibration models (especially PLS and PCR) have been extensively due to their ease of use, fast computation, good predictive performance, and easy interpretable representations [49]. However, the linearity assumption is not always valid, and when the spectra exhibit nonlinearities, they tend to give nonoptimal results [49]. Therefore, in such cases, it is of greater significance to develop the robust model system based on different nonlinear calibrations [49, 50]. These models include kernel PLS (KPLS), support vector machines (SVM), least squares SVM (LS-SVM), and among other artificial neural networks (ANN). Brief descriptions of these models are given in the subsequent paragraphs.

Kernel PLS is a nonlinear extension of linear PLS in which input data are transformed into a high-dimensional feature space via nonlinear mapping [51]. Briefly, the KPLS includes two steps. The first step includes embedding data in an input space via nonlinear mapping. The second step is that a linear algorithm is designed to discover the linear relationship [52]. The ANN on the other hand is a flexible mathematical structure capable of identifying complex nonlinear interactions between input and output data sets. This method is reported to be useful and efficient, especially in problems for which the characteristics are difficult to describe using physical equations [53]. The ANN model has been shown to perform better than linear models [54].

Support vector machines are the models that involve a solution of a quadratic programming problem leading to global models that are often unique [55]. The application of this type of model is further discussed and investigated by the authors in Refs. [56–59]. Finally, LS-VSM model is a simplification of the computational calculations of SVM by implementation of a least squares version for SVM [51]. Least squares SVM is capable of dealing with both linear and nonlinear multivariate calibration problems relatively fast [60]. In LS-VSM, a linear estimation is done in a kernel-induced feature space; the use of LS-SVM and NIR has been investigated by Borin et al. [55].

There are few studies on application of the nonlinear multivariate calibration in NIR spectroscopy that have been reported for analysis of trace metals. For instance, Shao and He [61] investigated the two sensitive wavelength (SW) selection methods combined with visible-near-infrared (vis/NIR) spectroscopy to determine the levels of some trace elements (Fe, Zn) in rice leaf. Calibration models using SWs selected by latent variables analysis (LVA) and independent component analysis (ICA) and nonlinear regression of a least squares support vector machine (LS-SVM) were developed. In the nonlinear models, six SWs selected by ICA provide the optimal ICA-LS-SVM model when compared with LV-LS-SVM. The coefficients of determination (R2), root-mean-square error of prediction (RMSEP), and bias by ICA-LS-SVM were 0.6189, 20.6510, and −12.1549 ppm, respectively, for Fe, and 0.6731, 5.5919, and 1.5232 ppm, respectively, for Zn [62]. The overall results indicated vis-NIR spectroscopy combined with ICA-LS-SVM provided accurate determination of trace elements in rice leaf. Other methods are reported by Xu et al. [62] and Barbosa et al. [63], among others.

3. Conclusions

This chapter revealed that NIRS combined with multivariate tools has a great potential tool to improve the understanding of trace metal concentration in environmental matrices. It was also concluded that the use of chemometrics offered a rapid and cost-effective alternative to measure multielement particularly in soils and sediments. However, according to literature chemometric models such as MLS and PCR used in the NIRS are not reliable as compared to PLSR. Therefore, it is important for researchers to select proper chemometrics for their application. Due to the limitations encountered when using some of the chemometric tools, it is necessary to develop new chemometric methods or modify the conventional ones so as to improve their reliability and accuracy. It is reported in the literature that the limitation of NIRS is that some trace metals and other mineral compounds such as phosphorus do not absorb radiation in the NIR region. However, in most cases, this problem is solved by using the absorption features of metals associated with molecular vibrations of organic and inorganic functional groups in organic matter, silicates, carbonates, and water in near infrared (780–2500 nm region). Alternatively, some researchers combine different detection techniques such as UV-visible with NIRS [17, 64].

Author details

Philiswa N. Nomngongo*, Tshimangadzo S. Munonde, Anele Mpupa and Nkositetile Raphael Biata

*Address all correspondence to: pnnomngongo@uj.ac.za

Department of Applied Chemistry, University of Johannesburg, Johannesburg, South Africa

References

[1] Wei, B., & Yang, L. (2010). A review of heavy metal contaminations in urban soils, urban road dusts and agricultural soils from China. Microchemical Journal, 94(2), 99–107.

[2] Kumar, P., & Singh, A. (2010). Cadmium toxicity in fish: an overview. GERF Bulletin of Biosciences, 1(1), 41–47.

[3] Obodai, E. A., Boamponsem, L. K., Adokoh, C. K., Essumang, D. K., Villawoe, B. O., Aheto, D. W., & Debrah, J. S. (2011). Concentrations of heavy metals in two Ghanaian Lagoons. Archives of Applied Science Research, 3, 177–187.

[4] Rajaganapathy, V., Xavier, F., Sreekumar, D., & Mandal, P. K. (2011). Heavy metal contamination in soil, water and fodder and their presence in livestock and products: a review. Journal of Environmental Science and Technology, 4(3), 234–249.

[5] Li, Z., Ma, Z., van der Kuijp, T. J., Yuan, Z., & Huang, L. (2014). A review of soil heavy metal pollution from mines in China: pollution and health risk assessment. Science of the Total Environment, 468, 843–853.

[6] El-Bouraie, M. M., El-Barbary, A. A., Yehia, M. M., & Motawea, E. A. (2010). Heavy metal concentrations in surface river water and bed sediments at Nile Delta in Egypt. Suo, 61(1), 1–12.

[7] Aderinola, O. J., Clarke, E. O., Olarinmoye, O. M., Kusemiju, V., & Anatekhai, M. A. (2009). Heavy metals in surface water, sediments, fish and perwinkles of Lagos Lagoon. American-Eurasian Journal of Agricultural & Environmental Sciences, 5(5), 609–617.

[8] Adelekan, B. A., & Abegunde, K. D. (2011). Heavy metals contamination of soil and groundwater at automobile mechanic villages in Ibadan, Nigeria. International Journal of Physical Sciences, 6(5), 1045–1058.

[9] Shaylor, H., McBride, M., Harrison, E. (2009). Sources and Impacts of Contaminants in Soil. Cornell Waste Management Institute, New York (USA), http://cwmi.css.cornell.edu accessed 15 May 2016.

[10] Dimpe, K. M., Ngila, J. C., Mabuba, N., & Nomngongo, P. N. (2014). Evaluation of sample preparation methods for the detection of total metal content using inductively coupled plasma optical emission spectrometry (ICP-OES) in wastewater and sludge. Physics and Chemistry of the Earth, Parts A/B/C, 76, 42–48.

[11] Guo, W., Zhang, P., & Jin, L. (2015). Heat-assisted slurry sampling GFAAS method for determination of lead in food standard reference materials. Journal of Food Composition and Analysis, 42, 78–83.

[12] Armstrong, K. C., Tatum, C. E., Dansby-Sparks, R. N., Chambers, J. Q., & Xue, Z. L. (2010). Individual and simultaneous determination of lead, cadmium, and zinc by anodic stripping voltammetry at a bismuth bulk electrode. Talanta, 82(2), 675–680.

[13] Raoof, J. B., Ojani, R., Alinezhad, A., & Rezaie, S. Z. (2010). Differential pulse anodic stripping voltammetry of silver(I) using p-isopropylcalix[6]arene modified carbon paste electrode. Monatshefte für Chemie-Chemical Monthly, 141(3), 279–284.

[14] Vanini, G., Souza, M. O., Carneiro, M. T., Filgueiras, P. R., Bruns, R. E., & Romão, W. (2015). Multivariate optimisation of ICP OES instrumental parameters for Pb/Ba/Sb measurement in gunshot residues. Microchemical Journal, 120, 58–63

[15] Moros, J., Llorca, I., Cervera, M. L., Pastor, A., Garrigues, S., & De la Guardia, M. (2008). Chemometric determination of arsenic and lead in untreated powdered red paprika by diffuse reflectance near-infrared spectroscopy. Analytica Chimica Acta, 613(2), 196–206.

[16] Moros, J., Martínez-Sánchez, M. J., Pérez-Sirvent, C., Garrigues, S., & de la Guardia, M. (2009). Testing of the Region of Murcia soils by near infrared diffuse reflectance spectroscopy and chemometrics. Talanta, 78(2), 388–398.

[17] Wang, J., Cui, L., Gao, W., Shi, T., Chen, Y., & Gao, Y. (2014). Prediction of low heavy metal concentrations in agricultural soils using visible and near-infrared reflectance spectroscopy. Geoderma, 216, 1–9.

[18] Shi, T., Chen, Y., Liu, Y., & Wu, G. (2014). Visible and near-infrared reflectance spectroscopy—an alternative for monitoring soil contamination by heavy metals. Journal of Hazardous Materials, 265, 166–176.

[19] Tripathy, S., Middha, A., Swain, S. R. (2016). Assessment of phytoconstituent of five faced *Elaeocarpus ganitrus* beads by FTIR and UV-VIS spectroscopic analysis. Imperial Journal of Interdisciplinary Research, 2(10).

[20] Yuan, X., Luo, K., Zhang, K., He, J., Zhao, Y., Yu, D. (2016). Combinatorial vibration-mode assignment for the FTIR spectrum of crystalline melamine: a strategic approach toward theoretical IR vibrational calculations of triazine-based compounds. The Journal of Physical Chemistry A, 120(38), 7427–7433.

[21] Reichert, A., Sussmann, R. (2016). The Zugspitze radiative closure experiment for quantifying water vapor absorption over the terrestrial and solar infrared–part 3: quantification of the mid-and near-infrared water vapor continuum in the 2500 to 7800 cm^{-1} spectral range under atmospheric conditions. Atmospheric Chemistry and Physics, 16(18), 11671–11686.

[22] Reeves, J. B., & Smith, D. B. (2009). The potential of mid-and near-infrared diffuse reflectance spectroscopy for determining major-and trace-element concentrations in soils from a geochemical survey of North America. Applied Geochemistry, 24(8), 1472–1481.

[23] O'rourke, S. M., Minasny, B., Holden, N. M., & McBratney, A. B. (2016). Synergistic use of Vis-NIR, MIR, and XRF spectroscopy for the determination of soil geochemistry. Soil Science Society of America Journal, 80(4), 888–899.

[24] Wu, D., He, Y., Shi, J., & Feng, S. (2009). Exploring near and midinfrared spectroscopy to predict trace iron and zinc contents in powdered milk. Journal of Agricultural and Food Chemistry, 57(5), 1697–1704.

[25] Song, S., Yuan, L., Zhang, X., Hayat, K., Chen, H., Liu, F., ... & Niu, Y. (2013). Rapid measuring and modelling flavour quality changes of oxidised chicken fat by electronic nose profiles through the partial least squares regression analysis. Food Chemistry, 141(4), 4278–4288.

[26] Reich, G. (2005). Near-infrared spectroscopy and imaging: basic principles and pharmaceutical applications. Advanced Drug Delivery Reviews, 57(8), 1109–1143.

[27] Chavez, P. F., Sacre, P. Y., De Bleye, C., Netchacovitch, L., Mantanus, J., Motte, H., Schubert, M., Hubert, P., & Ziemons, E. (2015). Active content determination of pharmaceutical tablets using near infrared spectroscopy as Process Analytical Technology tool. Talanta, 144, 1352–1359.

[28] Kleinebecker, T., Poelen, M. D., Smolders, A. J., Lamers, L. P., & Hölzel, N. (2013). Fast and inexpensive detection of total and extractable element concentrations in aquatic sediments using near-infrared reflectance spectroscopy (NIRS). PLoS One, 8(7), e70517.

[29] Shamsoddini, A., Raval, S., & Taplin, R. (2014). Spectroscopic analysis of soil metal contamination around a derelict mine site in the Blue Mountains, Australia. ISPRS Annals of the Photogrammetry, Remote Sensing and Spatial Information Sciences, 2(7), 75.

[30] Putra, A., Meilina, H., & Tsenkova, R. (2012). Use of near-infrared spectroscopy for determining the characterization metal ion in aqueous solution. In Proceedings of the Annual International Conference, Syiah Kuala University-Life Sciences & Engineering Chapter (Vol. 2, No. 2), Banda Aceh.

[31] Tsenkova, R., Kovacs, Z., & Kubota, Y. (2015). Aquaphotomics: near infrared spectroscopy and water states in biological systems. In Membrane Hydration (pp. 189–211). Springer International Publishing, Switzerland.

[32] Schwartz, G., Ben-Dor, E., & Eshel, G. (2012). Quantitative analysis of total petroleum hydrocarbons in soils: comparison between reflectance spectroscopy and solvent extraction by 3 certified laboratories. Applied and Environmental Soil Science, 2012, 1–11.

[33] Xian-Li, X. I. E., Xian-Zhang, P. A. N., & Bo, S. U. N. (2012). Visible and near-infrared diffuse reflectance spectroscopy for prediction of soil properties near a copper smelter. Pedosphere, 22(3), 351–366.

[34] Horta, A., Malone, B., Stockmann, U., Minasny, B., Bishop, T. F. A., McBratney, A. B., Pallasser, P., & Pozza, L. (2015). Potential of integrated field spectroscopy and spatial analysis for enhanced assessment of soil contamination: a prospective review. Geoderma, 241, 180–209.

[35] Huang, Z., Tao, W., Fang, J., Wei, X., & Du, Y. (2009). Multivariate calibration of on-line enrichment near-infrared (NIR) spectra and determination of trace lead in water. Chemometrics and Intelligent Laboratory Systems, 98(2), 195–200.

[36] Wu, Y. Z., Chen, J., Ji, J. F., Tian, Q. J., Wu, X. M. (2005). Feasibility of reflectance spectroscopy for the assessment of soil mercury contamination. Environmental Science & Technology, 39(3), 873–878.

[37] Soriano-Disla, J. M., Janik, L., McLaughlin, M. J., Forrester, S., Kirby, J., Reimann, C., & Team, T. E. G. P. (2013). The use of diffuse reflectance mid-infrared spectroscopy for the prediction of the concentration of chemical elements estimated by X-ray fluorescence in agricultural and grazing European soils. Applied Geochemistry, 29, 135–143.

[38] Hemmateenejad, B., Akhond, M., & Samari, F. (2007). A comparative study between PCR and PLS in simultaneous spectrophotometric determination of diphenylamine, aniline, and phenol: effect of wavelength selection. Spectrochimica Acta Part A: Molecular and Biomolecular Spectroscopy, 67(3), 958–965.

[39] Kemper, T., & Sommer, S. (2002). Estimate of heavy metal contamination in soils after a mining accident using reflectance spectroscopy. Environmental Science & Technology, 36(12), 2742–2747.

[40] Malley, D. F., Yesmin, L., Wray, D., & Edwards, S. (1999). Application of near-infrared spectroscopy in analysis of soil mineral nutrients. Communications in Soil Science & Plant Analysis, 30(7–8), 999–1012.

[41] Choe, E., Kim, K. W., Bang, S., Yoon, I. H., & Lee, K. Y. (2009). Qualitative analysis and mapping of heavy metals in an abandoned Au–Ag mine area using NIR spectroscopy. Environmental Geology, 58(3), 477–482.

[42] Chang, C. W., Laird, D. A., Mausbach, M. J., & Hurburgh, C. R. (2001). Near-infrared reflectance spectroscopy–principal components regression analyses of soil properties. Soil Science Society of America Journal, 65(2), 480–490.

[43] Ko, H. J., Choi, H. L., Park, H. S., & Lee, H. W. (2004). Prediction of heavy metal content in compost using near-infrared reflectance spectroscopy. Asian Australasian Journal of Animal Sciences, 17(12), 1736–1740.

[44] Dupuy, N., & Douay, F. (2001). Infrared and chemometrics study of the interaction between heavy metals and organic matter in soils. Spectrochimica Acta Part A: Molecular and Biomolecular Spectroscopy, 57(5), 1037–1047.

[45] Shi, T., Wang, J., Chen, Y., & Wu, G. (2016). Improving the prediction of arsenic contents in agricultural soils by combining the reflectance spectroscopy of soils and rice plants. International Journal of Applied Earth Observation and Geoinformation, 52, 95–103.

[46] Shi, T., Liu, H., Wang, J., Chen, Y., Fei, T., & Wu, G. (2014). Monitoring arsenic contamination in agricultural soils with reflectance spectroscopy of rice plants. Environmental Science & Technology, 48(11), 6264–6272.

[47] Shao, X., Bian, X., Liu, J., Zhang, M., & Cai, W. (2010). Multivariate calibration methods in near infrared spectroscopic analysis. Analytical Methods, 2(11), 1662–1666.

[48] Li, J., Zhang, Y., Cai, W., & Shao, X. (2011). Simultaneous determination of mercury, lead and cadmium ions in water using near-infrared spectroscopy with preconcentration by thiol-functionalized magnesium phyllosilicate clay. Talanta, 84(3), 679–683.

[49] Ni, W., Nørgaard, L., & Mørup, M. (2014). Non-linear calibration models for near infrared spectroscopy. Analytica Chimica Acta, 813, 1–14.

[50] Chen, Q., Cai, J., Wan, X., & Zhao, J. (2011). Application of linear/non-linear classification algorithms in discrimination of pork storage time using Fourier transform near infrared (FT-NIR) spectroscopy. LWT-Food Science and Technology, 44(10), 2053–2058.

[51] Gu, C., Xiang, B., Su, Y., & Xu, J. (2013). Near-infrared spectroscopy coupled with kernel partial least squares-discriminant analysis for rapid screening water containing malathion American Journal of Analytical Chemistry, 4(3), 111–116. doi: 10.4236/ajac.2013.43015.

[52] Rosipal, R., & Trejo, L. J. (2001). Kernel partial least squares regression in reproducing kernel Hilbert space. Journal of Machine Learning Research, 2(Dec), 97–123.

[53] Wang, K., Chi, G., Lau, R., & Chen, T. (2011). Multivariate calibration of near infrared spectroscopy in the presence of light scattering effect: a comparative study. Analytical Letters, 44(5), 824–836.

[54] Hsu, K. L., Gupta, H. V., & Sorooshian, S. (1995). Artificial neural network modeling of the rainfall-runoff process. Water Resources Research, 31(10), 2517–2530.

[55] Borin, A., Ferrao, M. F., Mello, C., Maretto, D. A., & Poppi, R. J. (2006). Least-squares support vector machines and near infrared spectroscopy for quantification of common adulterants in powdered milk. Analytica Chimica Acta, 579(1), 25–32.

[56] Zomer, S., Brereton, R. G., Carter, J. F., & Eckers, C. (2004). Support vector machines for the discrimination of analytical chemical data: application to the determination of tablet production by pyrolysis-gas chromatography-mass spectrometry. Analyst, 129(2), 175–181.

[57] Brudzewski, K., Osowski, S., & Markiewicz, T. (2004). Classification of milk by means of an electronic nose and SVM neural network. Sensors and Actuators B: Chemical, 98(2), 291–298.

[58] Fernandez Pierna, J. A., Volery, P., Besson, R., Baeten, V., & Dardenne, P. (2005). Classification of modified starches by Fourier transform infrared spectroscopy using support vector machines. Journal of Agricultural and Food Chemistry, 53(17), 6581–6585.

[59] Van Gestel, T., De Brabanter, J., De Moor, B., Vandewalle, J., Suykens, J. A. K., & Van Gestel, T. (2002). Least Squares Support Vector Machines World Scientific, Singapore, 2002.

[60] Suykens, J. A., & Vandewalle, J. (1999). Least squares support vector machine classifiers. Neural Processing Letters, 9(3), 293–300.

[61] Shao, Y., & He, Y. (2013). Visible/Near infrared spectroscopy and chemometrics for the prediction of trace element (Fe and Zn) levels in rice leaf. Sensors, 13(2), 1872–1883.

[62] Xu, L., Yan, S. M., Cai, C. B., & Yu, X. P. (2014). Nondestructive discrimination of lead (Pb) in preserved eggs (pidan) by near-infrared spectroscopy and chemometrics. Journal of Spectroscopy, 2014.

[63] Barbosa, R. M., Batista, B. L., Varrique, R. M., Coelho, V. A., Campiglia, A. D., & Barbosa, F. (2014). The use of advanced chemometric techniques and trace element levels for controlling the authenticity of organic coffee. Food Research International, 61, 246–251.

[64] Huang, G., Han, L., Yang, Z., & Wang, X. (2008). Evaluation of the nutrient metal content in Chinese animal manure compost using near infrared spectroscopy (NIRS). Bioresource Technology, 99(17), 8164–8169.